Airborne Particulate Matter

Pratap Kumar Padhy · Soumya Niyogi ·
Pulak Kumar Patra · Markus Hecker
Editors

Airborne Particulate Matter

Impact on Human Health and Environment

Springer

Editors
Pratap Kumar Padhy
Department of Environmental Studies
Visva-Bharati University
Santiniketan, West Bengal, India

Pulak Kumar Patra
Department of Environmental Studies
Visva-Bharati University
Santiniketan, West Bengal, India

Soumya Niyogi
Department of Biology and Toxicology
University of Saskatchewan
Saskatoon, SK, Canada

Markus Hecker
School of Environment and Sustainability
and Toxicology Centre
University of Saskatchewan
Saskatoon, SK, Canada

ISBN 978-3-031-84407-2 ISBN 978-3-031-84408-9 (eBook)
https://doi.org/10.1007/978-3-031-84408-9

This Springer imprint is published by the registered company Springer Nature Switzerland AG
The registered company address is: Gewerbestrasse 11, 6330 Cham, Switzerland

Preface

It is well-known that air pollution affects public health worldwide with significant economic burdens. Besides human health, air quality is closely linked to the global climate and ecosystem health. Economic development and rapid growth of industries and urbanization contribute to deteriorating air pollution. Thus, governments seek to improve air quality to lessen the burden and cost of air pollution-related public health issues. Research-based scientific knowledge supports formulating effective legal frameworks and policy measures to curb air pollution and mitigate its impacts. The academic field of 'air pollution and health' has expanded enormously in the last two decades. Realizing the complex issues in this field has diversified the research domains with a multidisciplinary approach. With our present understanding, control strategies and clean air legislations have addressed mostly tractable emissions sources such as vehicular and industrial pollution, significantly reducing sulphur dioxide and nitrogen oxides in the ambient air. However, a clear understanding of the mechanisms and pathways of air pollutants and their toxic effects is still lacking. The primary concerns are the non-exhaust particulate emissions, secondary organic aerosol (SOA), and sulfate and nitrate in the air. An in-depth analysis can provide implications for policymakers to adopt more stringent strategies to reduce air pollution and its health effects.

Airborne Particulate Matter (PM), consisting of a heterogeneous mixture of solid and liquid particles suspended in air, is the most potent pollutant. Airborne particles show great complexity because of their vast size and chemical composition variation over a range of temporal and spatial scales. Our understanding of these mechanisms and processes of transport and the impact of airborne particles has grown significantly, and there is consensus on the health hazards of PM. Still, disagreement exists over the precise mechanisms contributing to PM actions on human health and the environment. Thus, to establish the health effects of PM, the focus should be on their physicochemical characteristics and biochemical interactions with human cells and organs.

In this context, this edited volume addresses some critical issues related to air particulates and health, which have not been adequately addressed earlier. It provides an overview of current and upcoming research trends in monitoring, exposure studies, health and environmental risk assessment, and particulate air pollution mitigation measures. Appropriate case studies have been included to offer some insights into air pollution-related regional and local health and management issues. One of the characteristics of this book is the multidisciplinary approach that integrates the expertise of contributors from diverse research fields (environmental science, toxicology, geology, public health science, biology, physics, chemistry, and geospatial technology).

The book focusses on the effect of prolonged exposure to air pollution rather than the more familiar short-term health repercussions such as respiratory conditions. Air pollution studies have benefitted from technological advancement in other fields. The application of two such advanced technologies (isotope tracers and satellite-based geospatial technology) in monitoring and source characterization of particulate matter has been discussed at great length.

The health impacts and analytical techniques of two emerging air pollutants, i.e. oxygenated polycyclic aromatic hydrocarbons (PAHs) and microplastics (MPs), are described.

Conventionally, the health impact of air particulates is based on epidemiological and statistical studies. Establishing a more definite cause-and-effect relationship requires tracing the mechanism and clinical evidence. Thus, we have included chapters that discuss the exposure pathways and mechanisms of the health impact that link particulate matter to various respiratory, haematological, and neurological disorders. The mechanisms or pathways of PM and its constituent metals are discussed to explain their role in the cell's Reactive Oxygen Species (ROS) generation, accelerating the development of cardiovascular diseases and cancer. These findings may assist public health scientists in the early detection of air pollution impact.

Case studies from mining and urban regions of India have been included, focussing on developing predictive models from the observed data.

We have also included chapters dealing with PM's impact on crops and natural vegetation. Finally, a novel microbial-assisted phytoremediation technology is discussed as a mitigation strategy for air pollution.

We hope the book will provide readers with a scientific foundation for understanding particulate matter pollution and assist them in appreciating its role in public health. It also delineates future research initiatives and policy actions needed to protect human health from air pollution, locally and globally. The book will be of great educational value and recommended for consultation and teaching.

This volume results from an MHRD-sponsored SPARC project (Project Code: P1389) by the editors and an international conference on its agenda. We acknowledge the sponsoring agency. We are grateful to all the contributing authors for their contributions and patience. We also thank our publishers, Capital Publishing Company and Springer, Germany, for their timely support in bringing out the volume.

Santiniketan, India — Pratap Kumar Padhy
Saskatoon, Canada — Soumya Niyogi
Santiniketan, India — Pulak Kumar Patra
Saskatoon, Canada — Markus Hecker

Contents

About the Editors

Pratap Kumar Padhy Professor and Head at the Department of Environmental Studies, Institute of Science, Visva-Bharati, Santiniketan, India is working on various aspects of Air Pollution and Health and has research experience of more than 30 years. He has published many research papers in internationally refereed journals, published three edited books, and several chapters in books. Prof. Padhy has also undertaken several research projects, including one international collaborative project. He was awarded the President of India Gold Medal for General Proficiency from JNU, New Delhi during 1995–1996.

Soumya Niyogi Professor of Biology and Toxicology at the University of Saskatchewan, Saskatoon, Canada is also an adjunct professor of Biology at the York University, Toronto, Canada. He is an Environmental Toxicologist with over 25 years of research experience specializing in the toxicology of metals/metalloids and polyaromatic hydrocarbons (PAHs). He has undertaken many government and industry funded research projects and supervised many Ph.D. and M.Sc. students. He is on the editorial board of two reputed Toxicology journals.

Pulak Kumar Patra is an Associate Professor at the Department of Environmental Studies, Institute of Science, Visva-Bharati, Santiniketan. He has over 22 years of teaching and research experience in Environmental Science. His research interests include environmental geoscience, hydro-geochemistry and medical geology. He has published more than 50 research papers in journals and edited books. He has also authored three books and edited four books.

Markus Hecker is a Professor and Canada Research Chair at School of Environment and Sustainability, University of Saskatchewan. He is also a member of the College of the Royal Society of Canada, and a visiting/guest professor at the Peking Union Medical College and Xiamen University, China. His research interest includes

Ecotoxicology, Endocrine disruption and Environmental DNA. He is serving on the editorial board of many reputed journals. Markus has authored or co-authored over 170 peer-reviewed papers, review articles, editorials and book chapters. He is currently serving on the Board of Directors of the Society of Toxicology and Chemistry (North America).

Introduction

Pratap Kumar Padhy, Soumya Niyogi, Pulak Kumar Patra, and Markus Hecker

1 Introduction

Air pollution and its effect (or impact) on the environment and human health is of a grave concern. Almost all the people globally are believed to breathe air that exceeds WHO guideline limits. Annually, 8.9 million or 7.6% global deaths accounting for 103.1 million life years loss are caused by exposure to polluted air (Hassan et al., 2021). Airborne particulate matter (PM), consisting of heterogeneous mixture of solid and liquid particles suspended in air are more dangerous and complex than gaseous pollutants. Air pollution-induced health hazards impose a significant economic burden. Besides human health, air quality is closely linked to the global climate and ecosystem health. Thus, governments worldwide seek to improve air quality to lessen the burden and cost on air pollution related public health issues. The health risk assessment for air pollution supports policy decisions by estimating the health impacts expected from air quality measures in different situations. The number of deaths and years of healthy life lost worldwide has hardly decreased, despite some noteworthy improvements in air quality, particularly in more industrialised nations in the recent decades.

Economic development, with the rapid growth of industries and urbanisation in developing countries, including India, contributes to deteriorating air pollution (Ghosh et al., 2023). Research-based scientific knowledge supports formulating

P. K. Padhy (✉) · P. K. Patra
Department of Environmental Studies, Visva-Bharati, Santiniketan, Bolpur, West Bengal, India
e-mail: pkpadhy@visva-bharati.ac.in

S. Niyogi
Department of Biology, University of Saskatchewan, Saskatoon, SK, Canada

S. Niyogi · M. Hecker
Toxicology Centre, University of Saskatchewan, Saskatoon, SK, Canada

M. Hecker
School of Environment and Sustainability, University of Saskatchewan, Saskatoon, SK, Canada

P. K. Padhy et al. (eds.), *Airborne Particulate Matter*,
https://doi.org/10.1007/978-3-031-84408-9_1

effective legal frameworks and policy measures to curb air pollution and mitigate its impacts. The studies in air pollution and health have progressed and diversified exponentially in the last few decades. These findings help set-up air quality standards and implement air pollution management/legislation. This volume provides an overview of current and upcoming research trends in monitoring, exposure studies, health and environmental risk assessment, and mitigation measures of particulate air pollution. It includs contributions from research fields following a multidisciplinary and integrated approach.

2 Nature and Source of Air Particulate

Airborne particles show great complexity because of their wide variation in size and chemical composition over a range of temporal and spatial scales. This is in contrast to atmospheric trace gases, which are uniform in their physical and chemical properties wherever they exist. The physicochemical characteristics of PM are vital factors that influence their health impacts. The size distribution of airborne particles varies widely with their diameter ranging from few nanometres to tens of micrometres. Fine fraction particles are generally produced by high-temperature processes like automobile exhaust or industrial combustion, or by gas-to-particle conversion reactions occurring in the atmosphere.

Coarse fraction particles, on the other hand, are associated with mechanical wearing and tearing processes such as abrasion of bedrocks and other materials, resuspension of soil and fine particles by aeolian or formation of sea spray through wave breakings. The largest airborne particles are more than 100 μm in diameter but have a relatively short atmospheric lifetime due to high gravitational settling speeds. The air quality standards and guidelines for PM use two health-relevant fractions, i.e., PM10 for particles smaller than 10 μm and PM2.5, for particles smaller than 2.5 μm. Particles whose at least one dimension is smaller than 100 nm (0.1 μm) are called ultrafine or nanoparticles. Their suspected toxicity and regional nucleation processes has contributed to the current interest in them. The particulate size distribution reflects predominant sources of emission and the particle growth mechanism in the atmosphere. Particles grow faster in more polluted atmospheres through coagulation, which is dependent on particle number concentrations, and condensational growth, which is mostly caused by atmospheric oxidation processes. This results in low volatility species that condense onto pre-existing particles and accelerate their growth. Besides size, the morphology of particulates is also an essential factor to reckon with.

Chemically, airborne particulates include both organic and inorganic constituents. They may be of primary particles (directly emitted from anthropogenic and natural sources) or secondary particles (formed through chemical reactions among gaseous and particulate matter in the atmosphere). Organic carbon (OC) includes biological compounds such as cell fragments or materials of biological origin such as polycyclic aromatic hydrocarbons. Secondary organic aerosol (SOA) that includes volatile organics are the most complex in nature due to involvement of diverse chemical reactions. The inorganic carbonaceous material is classified into black carbon (BC) and elemental carbon (EC). Other inorganic constituents are metals including potential toxic elements such as lead, iron, copper, nickel, vanadium, chromium and silicon. Sulphates and nitrates derived from their gaseous oxides are also found in variable proportions. Some components of the PM mixture may be of more significant public health concern than others. While the PM over the ocean is more or less homogenous, containing mainly sea salt and oxidation products of trace gases from the oceans, PM over the land shows significant local and temporal variation with a combination of primary and secondary components derived from diverse natural and anthropogenic sources. The PM10 fractions mainly consist of insoluble crust-derived minerals, biological material (such as pollen, endotoxins, fungi, and bacteria), and sea salts. By contrast, PM2.5, including the ultrafine fractions, is derived mainly from combustion-related sources (Harrison, 2020). Mathematical models with multidisciplinary approaches have been employed to infer the source of airborne particles. Data related to known emissions sources and disperse patterns have been traditionally used in dispersion and chemistry-transport modelling. It is now being complemented with the in-receptor modelling that use air quality data (Almeida et al., 2020).

Black carbon released from burning of coal has been the major contributor to particulate pollution in the 19th century. However, since second half of the 20th century, secondary pollutants and carbonaceous particles from vehicle exhaust gradually became dominant. Control strategies are now addressing the more tractable emissions sources such as vehicular and industrial pollution. However, non-exhaust particulate emissions such as surface abrasion particles, tyre dust, brake abrasion particles, dust from the attrition of the disc and pad, and resuspended road surface dust are increasing as there is no legislative control. Another emerging pollutant is secondary organic aerosol (SOA) whose concentration has not been affected by reductions in VOC emissions. Similarly, the decrease in emissions of sulphur dioxide and nitrogen oxides over the past decades, has successfully reduced their concentrations in the air. However, their derivatives, e.g. sulphate and nitrate, have not fallen proportionately.

3 Health Risk Assessment

Understanding these mechanisms and processes of transport and impact of airborne particles has grown significantly over the past few decades and there is consensus on health hazard of PM. Still disagreement exists over the precise mechanisms that contributes to PM actions on human health and environment. Thus, to establish the health effects of PM, focus should be on their physicochemical characteristics and biochemical interactions with human cells and organs. Numerous processes, such as population exposure to pollutants, and the evaluation of adverse health impacts using particular concentration-response functions, are mathematically estimated and modelled in the health risk assessment for air pollution.

Both epidemiologic and toxicologic methods have been used in several studies conducted over the decades to assess the toxicity of PM constituents. Controlled exposure research, in nonhuman animals has shown that the solubility, physicochemical makeup, and sizes of the particles influence their harmful effects.

The outcomes of air pollution health risk assessment are frequently condensed into a variety of metrics, such as the number of patients or deaths, years of life lost (YLL), disability-adjusted life years (DALY), or variations in life expectancy. The data on health impacts, community risk, exposure to air pollution is analysed for regulatory decision-making and public engagement. Local topography, weather conditions, seasons, sources and concentration of particle influence the exposure effectiveness of PM. The vulnerability of a person to PM exposure depends on the physical characteristics of its breathing (volume, rate and mode). Particle size is also playing a key role. The bulk of coarse PM (particles larger than 10 μm in diameter) are efficiently filtered during inhalation by the nasal cavity's cilia and mucus. According to Atkinson et al. (2010), the trachea, or upper throat, is where the coarse PM portion settles. If we breathe in this PM, it will first congregate in our nose and throat. Our body will then eliminate these intrusive PMs through reactions like sneezing and coughing (Cadelis et al., 2014). Smaller particles generally enter the respiratory system deeper and deposit on it more quickly. They can pass through the nasal passages and reach the alveoli in the respiratory system. The tracheobronchial tree is most likely where particles between 5 and 10 μm are deposited. On the other hand, particles smaller than 5 μm are found in the alveoli and respiratory bronchioles, which are the sites of gas exchange. They can enter the lung and affect the gas exchange in the lungs. Eventually, they contaminate the bloodstream resulting in serious health issues. Particles smaller than 1 μm go further into the circulatory system and cell tissues by diffusion forces and behave like gas molecules (Valavanidis et al., 2008). Transition metals (particularly iron) found in particles enhance the generation of reactive oxygen species (ROS) which may damage cells and tissue initiating or aggravating inflammation (Kim et al., 2015).

The majority of the evidence regarding the close relationships between PM and health comes from epidemiological studies. This indicates that while the statistical evidence supports the association, the inference is oblique and fails to establish a mechanical relationship. Recent research in this area has begun to address the issue combining epidemiological data with PM properties to maximise the association between health effects and diseases. Despite the vast and growing body of data from *in vitro* and *in vivo* experiments, identification of the responsible chemical species and understanding their precise role remains a challenge.

4 Design of the Book

With contributions of researchers from diverse fields and a multidisciplinary approach, the book provides an overview of the recent trends and modern research techniques in the field of airborne particulate matter and their impact on human health and the environment. The thirteen chapters deal with the emerging trends and challenges in monitoring, source characterisation, health risks assessment, environmental impacts as well as novel management strategies.

Chapter 2 contributed by eminent medical professionals, outlines air pollution impact on human health. The authors distinguish the effect of prolonged exposure to air pollution from the more familiar short-term health repercussions such as respiratory conditions including asthma and bronchitis and eye, nose and skin irritation. Long-term health effects such as chronic respiratory, cardiovascular, and neurological conditions are more challenging and difficult to establish the cause-and-effect linkage. Young children, older people and people with pre-existing medical issues are most vulnerable. Air pollution contributes to climate change, enhancing the risk of infectious diseases. The authors believe addressing this multifaceted problem requires raising public awareness and adopting an interdisciplinary strategy. In addition to managing the adverse health effects of air pollution, developing a global prevention policy to combat anthropogenic air pollution is essential. Sustainable development practices and utilization of research-based knowledge can address the issues effectively.

The next two chapters deal with the application of two advanced technologies in monitoring air particulates. Chapter 3 shows that isotope tracers are becoming increasingly popular for source identification of air particulates and the development of predictive models in atmospheric chemistry. It provides a brief synopsis of the background information on using isotope tracers in air pollution studies. This study underscores the significance of the interconnectedness between the hydrosphere, atmosphere, and lithosphere in shaping the isotopic composition, hence enabling the identification of many variables that contribute to air pollution. The application of isotopic ratios as source tracers depends on the fact that the isotopic composition of the pollutants undergoes little change during transportation and isotopic differences exist among the various sources. The chapter also deals with analytical procedures, calculations, and result interpretation using isotope technologies to identify pollution sources. Chapter 4 discusses the spaceborne remote sensing methods which have been widely employed to retrieve information on various air pollutants, especially particulate matter. Authors identified a growing need for integrated air quality data from remote sensing satellites and ground-based monitoring stations for health effect evaluations. The limitations and the future trends of the RS-GIS method in interpreting the results of the air pollution and health effects research are also discussed.

Chapters 5 and 6 deal with two emerging air pollutants, i.e. oxygenated polycyclic aromatic hydrocarbons (PAHs) and microplastics (MPs), respectively. The incomplete combustion of organic materials, such as coal, oil, petrol and wood, results in the production of polycyclic aromatic hydrocarbons (PAHs), which are common environmental contaminants. There is growing interest in the chemical analysis, occurrence, fate and behaviour of oxygenated PAHs due to strong indication of their toxicological importance. Chapter 5 focusses on state-of-the-art innovations in the chemical analysis of oxygenated PAHs with focus on their significant chemical-physical properties, primary sources and routes to the atmosphere, and health effects. The sample analysis techniques have been discussed along with their collection, preparation and separation. Microplastics pollution (MPs) has led to the global problem due to the careless disposal of plastic trash. Although earlier study on MPs and their effects was limited to terrestrial and aquatic environments, more recent studies have examined MPs in the atmosphere. Their effects on atmospheric transport to pristine habitats and urban air quality have become a major concern. Nonetheless, little is known about the effects of MPs on human health. Chapter 6 documents the prevalence of microplastic burden in lung tissue from lung cancer patients and summarises the current state of MP research. It highlights the lack of information and gaps in our understanding of atmospheric movement and the role played by particle plastics.

Most studies on the health impact of air particulates are based on epidemiological and statistical studies. The focus has been now shifted is to trace the mechanism and clinical evidence to establish a more definite cause-and-effect relationship. Chapters 7, 8 and 9 discuss the exposure pathways and mechanism of health impact. Chapter 7 links air pollutants to various respiratory and haematological disorders. It reviews the effects of PM on platelet, white blood, and red blood cell activities. Haematological disorders may be developed due to PM2.5 exposure in concert with several other risk factors. More research in this field can be helpful in early detection of some diseases. Chapter 8 provides an overview of the impact of particulate matters on neurological disorders. Relationship between particulate matter and neurological disorders or neurodevelopmental disturbances has been established through observational, epidemiological, experimental and clinical studies. Air pollution causes debilitating central nervous system (CNS) disorders, including an increased risk of stroke, decline in cognitive function, Parkinson's and Alzheimer's diseases. Air pollution induces neurological inflammation, oxidative damage and neurodegenerative illness. Mechanically, air particles impact on the central nervous system (CNS) through a variety of molecular, cellular, and inflammatory processes that can cause direct harm to brain tissue or result in neurological diseases.

Oxidative stress due to exposure to PM accelerates the development of cardiovascular diseases and cancer. Chapter 9 traces the mechanisms or pathways of PM and its constituent metals to explain their role in reactive oxygen species (ROS) generation in the cell. The cellular (dichlorofluorescein, MTT and ascorbic test) and acellular (dithiothreitol, glutathione, ascorbic acid, and electron spin resonance technique) assays used for determining the oxidative potential of airborne particulate matter are discussed.

Chapters 10 and 11 are case studies from different regions of India focussing on development of predictive models from the observed data. Chapter 10, attempts to predict the seasonal variation of PM10 and PM2.5 in Singrauli coal mines, Madhya Pradesh, India using air quality modelling software "AERMOD" (American Meteorological Society/Environmental Protection Agency Regulatory Model). The prediction of SPM before the excavation helps reduce the risk of environmental impact and human health. The open-cast mine contributes a large amount of RSPM in the surrounding area. The yearly meteorological data and the emission rate calculated for haulage routes, railway siding, drilling, and dumping sites are used as input parameters to predict the model. This research will help to enhance the applicability of AERMOD in Indian geo-mining conditions. Chapter 11 provides a statistical prediction for the PM2.5 health risk plot from 2016 to 2025 in Bhubaneswar, Odisha (India) and revealed a disrupted trend caused by sudden suppressed sources of emissions during COVID-19 epidemic. It considers seasonal health risk assessment with a gender bias in exposure. Though the exposure pattern for both males and females followed the same seasonal trend, proportional to the PM concentration level with high exposure dose during the winter season and lowest during monsoon, male exposure doses were consistently higher than female exposure doses for each exposure condition. The authors also showed that individual health risks are reduced by one or two orders of magnitude by wearing masks.

Chapters 12, 13 and 14 deal with the impact of PM on the vegetation. Particulate matter adversely affects the leaves of the plants, resulting in damage to chlorophyll, carotenoids, and ascorbic acid contents and decrease the plant productivity. Chapter 12 is based on the study of effect of automobile pollution on morphological and biochemical parameters of four tree plants (*Mangifera indica, Ficus religiosa, Polyalthia longifolia* and *Butea monosperma)* in Chapra town of Bihar, India. The study showed that change in photosynthetic pigments is directly associated with the levels of air pollution and *Ficus religiosa* has a high air pollution tolerance index.

In Chapter 13, authors have discussed the role of PM as a significant environmental stressor affecting rice crops and contributing to GHG emissions. PM deposition on rice plants can have multifaceted and complex effects on their physiology and yield. The extent of these effects depends on factors such as PM concentration, particle size, composition, rice cultivation, and environmental conditions. Understanding these interactions is essential for developing effective strategies to mitigate the negative impacts of PM deposition on rice crops and enhance agricultural sustainability.

Chapter 14 showcases the potential use of plants in decontaminating heavy metal-polluted areas, including atmospheric pollution. Root-associated microbes of the bioremediating plant(s) might have a role in the conversion of oxidation states of heavy metals, thereby helping the plant(s) accumulate HMs in their respective forms. Authors think that bioremediation through conventional plants alone is not effective. This can be solved by microbial-assisted phytoremediation combining genetic engineering and biochar supplements.

We hope the book will provide readers a scientific foundation for understanding PM pollution and assist them in appreciating its role in public health. It will promote further study and legislative actions to curb the pollution and improve the quality of life.

References

Almeida SM, Manousakas M, Diapouli E, Kertesz Z, Samek L, Hristova E, Šega K, Alvarez RP, Belis CA, Eleftheriadis K, IAEA European Region Study GROUP (2020). Ambient particulate matter source apportionment using receptor modelling in European and Central Asia urban areas. Environ Pollut 266:115199

Cadelis G, Tourres R, Molinie J (2014) Short-term effects of the particulate pollutants contained in saharan dust on the visits of children to the emergency department due to asthmatic conditions in Guadeloupe (French Archipelago of the Caribbean). PLoS ONE 9(3):e91136. https://doi.org/10.1371/journal.pone.0091136

Ghosh B, Padhy PK, Niyogi S, Patra PK, Hecker MA (2023) Comparative study of heavy metal pollution in ambient air and the health risks assessment in industrial, urban and semi-urban areas of West Bengal, India: an evaluation of carcinogenic, non-carcinogenic, and additional lifetime cancer cases. Environments 10:190. https://doi.org/10.3390/environments10110190

Harrison RM (2020) Airborne particulate matter. Philos Trans R Soc A 378(2183):20190319

Hassan Bhatt T, Jiawen G, Farzaneh H (2021) Air pollution health risk assessment (AP-HRA), principles and applications. Int J Environ Res Public Health 18(4):1935. https://doi.org/10.3390/ijerph18041935

Kim KH, Kabir E, Kabir S (2015) A review on the human health impact of airborne particulate matter. Environ Int 74:136–143. https://doi.org/10.1016/j.envint.2014.10.005

Valavanidis A, Fiotakis K, Vlachogianni T (2008) Airborne particulate matter and human health: toxicological assessment and importance of size and composition of particles for oxidative damage and carcinogenic mechanisms. J Environ Sci Health C Environ Carcinog Ecotoxicol Rev 26(4):339–362. https://doi.org/10.1080/10590500802494538

Air Pollution and Human Health

Shambo S. Samajdar and Saibal Moitra

1 Introduction

Global disparities in exposure to air pollution are growing, especially as low- and middle-income nations encounter rising levels of air pollution as a result of pervasive urbanization and economic growth that has mainly relied on the combustion of fossil fuels. The World Health Organization estimates that air pollution annually causes millions of fatalities, primarily from noncommunicable diseases. Clean air should be considered a basic human right and a precondition for a healthy and productive existence. Over the past three decades, there have been modest improvements in air quality, but millions of people still die early, frequently from the most marginalized populations. The world's assessments of atmospheric air pollution alone indicate loads of thousands and thousands of wholesome lifestyles years misplaced, with low- and middle-earnings countries bearing the best disorder burden. People with continual conditions (which includes asthma, continual obstructive pulmonary disorder, and cardiovascular sickness), as well as elderly humans, babies, and pregnant girls, are particularly prone to the bad health results of air pollutants. In 2019, more than 90% of the world's population resided in areas in which $PM_{2.5}$ concentrations handed the 2005 WHO guideline for prolonged exposure.

It is estimated that air pollution causes 7 million deaths annually and much higher morbidities. In adults, ischemic coronary heart sickness and stroke are the leading causes of untimely mortality attributable to outdoor air pollution, and evidence is emerging of different outcomes, consisting of diabetes and neurodegenerative disorders. As a result, the amount of disease linked to air pollution is now comparable

S. S. Samajdar
Department of Clinical Pharmacology, School of Tropical Medicine, Kolkata, West Bengal, India

S. Moitra (✉)
Division of Allergy & Immunology, Apollo Multispeciality Hospital, Kolkata, India
e-mail: saibal.moitra@gmail.com

P. K. Padhy et al. (eds.), *Airborne Particulate Matter*,
https://doi.org/10.1007/978-3-031-84408-9_2

to other major global health risks including poor nutrition and smoking. Air pollution is one of the biggest environmental risks to human health, along with climate change. Air quality may be improved while lowering emissions, which can help with attempts to slow down climate change via striving to achieve these guiding principle levels, international locations will concurrently guard public health and mitigate climate change. The reduction of these so-called "classical" pollutants, such as particulate matter (PM), ozone (O_3), nitrogen dioxide (NO_2), sulphur dioxide (SO_2), and carbon monoxide (CO), has an impact on other dangerous pollutants. The risk to public health posed by particulate matter with diameters of 10 microns or less and 2.5 (PM_{10} and $PM_{2.5}$, respectively) microns are particularly significant. While both $PM_{2.5}$ and PM_{10} may penetrate far into the airways, only $PM_{2.5}$ can reach the bloodstream, where it can have an impact on several organs, particularly the heart and lungs. PM is predominantly produced by the combustion of fuels in a variety of industries, including transportation, energy, households, industry, and agriculture. In 2013, the WHO's International Agency for Research on Cancer (IARC) classified particle matter and outdoor air pollution as carcinogenic. Insufficient quantitative evidence is currently available to establish air quality guideline levels for nice varieties of PM (which encompass black carbon/elemental carbon, extraordinarily-wonderful debris, and debris originating from sand and dust storms). They are universally applicable in both outdoor and indoor settings (Goshua et al., 2022).

2 Sources of Air Pollution

It is recognized that majority of the environmental pollutants are released by large-scale anthropogenic activities, i.e. usage of industrial equipment, energy-producing facilities, fuel combustion and cars. Due to the scale at which these activities are performed, they are by far the biggest sources of air pollution, with motors thought to be responsible for 80% of the pollution that is present today (Moller et al., 1994). Fewer natural resources, such as volcano eruptions and forest fires, have an influence on our environment than other human activities, such as subject farming practices, gas stations, gas tank warmers and cleaning methods (Jacobson and Jacobson, 2002). Based on the origins of the pollutants, the air pollution is categorized. In accordance with the classification gadget, it is vital to say the four primary sources are fundamental assets, area sources, mobile sources, and herbal resources.

The primary sources of air pollutants include power plants, refineries, and petrochemical plants, as well as the chemical and fertilizer industries, the metallurgical and other industry plants, and, ultimately, municipal incineration. Location assets consist of residential cleaning, dry cleansing, printing industries and refueling stations. Motors, trains, airlines, and different vehicles are mobile sources. As indicated formerly, natural resources additionally consist of bodily calamities like forest fires, volcanic erosion, dust storms, and agricultural burning.

Though several class schemes have been proposed, a second form of categorization is based on the recipient of pollutants. Air pollutants is the presence of pollutants in widespread portions and for prolonged periods of time. Dispersed particles, hydrocarbons, CO, NO, NO_2, SO_2, and many other substances contribute to air pollution. Water contamination is the result of high levels of organic, inorganic, and biological charges. Soil pollution is caused by chemical releases or waste disposal, including pesticides, hydrocarbons, and heavy metals. By exploiting contaminating precipitation that falls into soil and aquatic settings, air pollutants can have a significant negative influence on soil and water. Extensive acid precipitation can alter the soil's chemistry via affecting plant life, cultures, and water fine. Similarly, soil acidity facilitates the motion of heavy metals, which then migrate into the aquatic surroundings. Heavy metals along with aluminum are recognised to be poisonous to fauna and animals. The importance of soil high-quality as soils with low calcium carbonate levels are more vulnerable to acid rain. Similarly, snow and particles seep into water bodies.

Pollutants are ultimately categorised based on their source, including radioactive and nuclear pollution, which include the discharge of radioactive and nuclear contaminants into the water, air and soil, as a result of nuclear explosions and accidents, nuclear weapons, and the management or disposal of radioactive sewage. Radioactive chemicals might pollute groundwater, humans, plants and animals since they are hazardous to the environment. Radium and uranium are two radioactive chemicals that are known to accumulate in the bones and cause cancer. Machinery, motors, tourist noise, and musical installations all emit noise pollution that are harmful.

Disability-adjusted life years (DALYs) were established by the global fitness industry. Years lost due to disability (YLD) for those who live with fitness condition or its effects, as well as years lost due to early death in the population (YLL). Air pollution, which is detected by noise pollutants, is the main cause of DALYs lost in Europe. Air pollution, noise and human health are related with each other. Given that ambient noise's effects on cardiovascular disease are independent of air pollution, it should be considered a legitimate public health concern (Stansfeld, 2015).

Adjustments inside the physical, chemical, or organic additives of the environment (air quality, temperature, weather, and so on) represent environmental pollutants. Pollutants impair the surroundings through both rising degrees above every day or with the aid of introducing noxious materials that are dangerous. Number one pollution is produced directly via the aforementioned sources, even as secondary pollution are byproducts of the number one pollution. As previously said, pollution can be natural or man-made, biodegradable or non-biodegradable. Similar to that, its basis might consist of a single source (factor-supply) or a variety of sources. The discrepancy in their ability to produce hazardous consequences can be explained by extraordinary physical and chemical sources of pollution. Aerosol chemicals are more harmful than gaseous compounds due to their minute size (solid or liquid) in the environment and higher capacity to penetrate. Our respiration device gets rid of gaseous compounds extra with ease. Those particles can damage the lungs and might even infiltrate the bloodstream, resulting inside the premature mortality of tens of millions of humans every year (Ethical, 2019). In addition, the acidity of aerosols (H^+) appears to seriously boom the manufacturing of secondary organic aerosols (SOA).

3 Climate Change and Air Pollution

Air pollution and climate change are closely connected. The variable that has the most impact on our planet's comfort is the climate. Aerosols, methane, tropospheric ozone, black carbon, and methane are pollutants that change the amount of incoming sunlight. As a result, as the earth's temperature rises, ice, icebergs and glaciers melt. The incidence and prevalence of imported and residual illnesses in Europe will be impacted by climate changes in a similar manner (Bezirtzoglou et al., 2011). Weather alter the global spread of infectious diseases and have a significant impact on the length, timing, and depth of epidemics. Climate change is extremely sensitive to parasitic or viral diseases spread by mosquitoes, as warming first shortens the pathogen's incubation period and then changes the geographic distribution of the vector. Additionally, the incidence of waterborne diseases will rise as a result of water warming by meteorological changes. Eliminated diseases including malaria, cholera, poliomyelitis, and tick-borne encephalitis appear to have lately reemerged in Europe as a result of population mobility. The spread of diseases is linked to cyclones and weather-related natural disasters, which now seems to be happening more frequently. Malnutrition and immune system imbalance are also linked to rising illnesses that are impacting public health. Due to outbreaks of disease reported in France and Italy, the Chikungunya virus "took the aeroplane" from the Indian Ocean to Europe (Calba et al., 2017).

3.1 Air Pollutants

The lead, sulphur oxide, nitrogen oxide, ground-level ozone, and particle pollution are the six main air pollutants that the WHO provides information on. A few of the environmental components that air pollution may have terrible effects on are groundwater, sediment and air. Additionally, it poses a grave threat to all living beings. As a result of their more severe and ubiquitous impacts on both human health and the environment, these pollutants are the major focus of research. The greenhouse effect, acid rain, climate change, and global warming all have significant ecological consequences on air pollution.

4 Particulate Matter (PM) and Health

Studies focussing on either brief-term (acute) or long-term (chronic) PM exposure have found a link between PM and detrimental health effects. PM is frequently created within the environment by the chemical interactions of various pollutants. Debris' capacity to penetrate is inversely correlated with its size. PM was defined by the American Environmental Protection Agency as a time period for debris. PM_{10}, or trash with a diameter of 10 μm or less, and $PM_{2.5}$, or particles with a diameter of typically 2.5 microns or less, make up PM pollutants. Tiny liquid or solid particles that can be inhaled to cause serious health effects are called particulate matter. PM_{10} particles can reach the lungs and even the bloodstream after being breathed. However, as given in Table 1, $PM_{2.5}$ particles represent a bigger threat to human health. The effects of PM on health have been the subject of several epidemiological research. Table 1 shows several investigations have identified a link between acute nasopharyngitis and both short- and long-term exposure to $PM_{2.5}$. Additionally, chronic exposure to PM has been linked to newborn and cardiovascular death.

Due to their dependence on $PM_{2.5}$ monitors, which restricts their reach to certain study regions or cities with a dearth of spatially resolved daily $PM_{2.5}$ concentration data, it is crucial to keep in mind that these studies could not accurately reflect the total population. The relative magnitudes of short- and long-term effects are not yet fully understood, according to a recent epidemiological study by the Department of Environmental Health at Harvard School of Public Health (Boston, MA) (Kloog et al., 2013). Spatial variations in $PM_{2.5}$ concentrations have been linked to an exposure error known as the Berkson error. The study team created a $PM_{2.5}$ exposure model based on remote sensing data to solve this problem. This model enables the geographical resolution of short-term impacts and the assessment of long-term consequences. The long-term chronic effects of PM exposure can include respiratory illnesses and immune system issues. It is important to emphasise that those with respiratory and cardiovascular disorders, diabetes, asthma, pneumonia, and other underlying medical conditions are especially prone to the negative effects of PM. Due to their tiny size, which allows them to enter indoor areas, $PM_{2.5}$ and PM_{10} show high connections with a variety of respiratory system illnesses (Boschi, 2012). Based on their chemical and physical characteristics, these particles have harmful impacts. Organic substances such polycyclic aromatic hydrocarbons, dioxins, benzene, and 1-3 butadiene, as well as inorganic substances like carbon, chlorides, nitrates, sulphates, and metals, can be found in PM_{10} and $PM_{2.5}$.

PM is divided into four major categories according to their nature and size (Heal et al., 2012) (Table 2). Different kinds of airborne PM are included in gas pollutants.

Table 1: Penetrability according to size of particle

Particle Size (μm)	*Penetration into human respiratory system*
>11	Passage into nostrils and upper respiratory tract
7-11	Passage into nasal cavity
4.7-7	Passage into larynx
3.3-4.7	Passage into trachea-bronchial area
2.1-3.3	Secondary bronchial area passage
1.1-2.1	Terminal bronchial area passage
0.65-1.1	Bronchioles penetrability
0.43-0.65	Alveolar penetrability

Table 2: Types and sizes of PM

Type	*Sub-types*	*PM diameter (μm)*
Particulate contaminants	Soot	0.01-0.8
	Smog	0.01-1
	Cigarette smoke	0.01-1
	Fly ash	1-100
	Cement dust particles	8-100
Biological contaminants	Viruses	0.01-1
	Allergens (dogs, cats, pollen, household dust)	0.1-100
	Bacteria and bacterial spores	0.7-10
	Fungi and molds	2-12
Types of dust	Settling dust	1-100
	Heavy dust	100-1000
	Atmospheric dust	0.01-1
Gases	Different gaseous contaminants	0.0001-0.01

Examples of particulate pollutants include smoke, soot, cigarette smoke, hydrocarbon smoke, fly ash and cement dust. In addition, biological pollutants such as cat allergies, home dust allergens, pollen, and microorganisms (bacteria, viruses, fungus, mould and bacterial spores) can be detected in the air. Dust comprises heavier particles, dust that settles, and suspended air dust. Due to their tiny size, PM_{10} and $PM_{2.5}$ particles have a longer lifetime in the atmosphere. This makes it possible for them to travel more easily and stay suspended in the air for longer time, subjecting both people and the environment to the same degree of pollution (Wilson and Suh, 1997). They could alter the balance of nutrients in aquatic habitats, harm forestry and agriculture, and acidify water supplies. As previously stated, $PM_{2.5}$ particles' tiny size contributes to more severe health effects. The main reason why "haze" forms in many urban areas is because of such little particles.

5 Air Pollutants and Environment

5.1 Ozone (O_3)

Ozone is a gas generated through the electrical discharge of oxygen at high voltage. It is 52% stronger than chlorine as an oxidant. While it originates in the stratosphere, ozone can also be produced as a result of photochemical pollution chain reactions in the troposphere. Being transported by air masses, ozone can travel long distances from its initial source. Notably, ozone levels above metropolitan areas are higher than those in cities, which might endanger cultures, forests, and plants by limiting carbon uptake. Ozone inhibits plant development and yield and has antibacterial effects on the microflora of plants. It also has an impact on several natural environments, changing the species makeup of both microflora and animals. Ozone's negative impacts include reduced cellular activity and DNA damage to epidermal keratinocytes. Nitrogen oxides and volatile organic compounds (VOCs), which may be released by natural sources or as a result of human activity, interact chemically to produce ground-level ozone (GLO). Ozone primarily affects the outer layers of the epidermis and tear ducts when it is inhaled. Mice exposed to high amounts of ozone for a brief period developed malondialdehyde in their epidermis and lost vitamin C and E. It is unclear, nevertheless, if elevated ozone levels affect how well and how consistently the skin barrier works, hence raising the risk of skin illness. Because ozone cannot dissolve in water, it can enter the alveoli deeply when inhaled. In metropolitan cities, ozone causes biochemical, morphological, functional, and immunological issues. Investigating the immediate impact of ambient ozone levels on mortality is the main goal of the European project known as APHEA2. Several European cities reported daily ozone readings and the number of fatalities over a 3-year period. Increase in ozone concentration during the mild season was associated with an increase in the number of fatalities (0.33%), respiratory deaths (1.13%), and cardiovascular deaths (0.45%). During winter, no impact was seen (Gryparis et al., 2004).

5.2 Carbon Monoxide (CO)

The incomplete combustion of fossil fuels results in the production of carbon monoxide. Headache, vertigo, weakness, nausea, vomiting, and finally loss of consciousness are some of the symptoms of carbon monoxide poisoning. It has negative consequences because it has a far stronger affinity for haemoglobin than for oxygen. Severe poisoning can result from prolonged exposure to high levels of carbon monoxide. Carbon monoxide can result in cardiovascular disorders, ischemia, and hypoxia because of its competitive binding with oxygen. It also contributes to the greenhouse effect, which is directly related to global warming and climate change. This may cause increased soil and water temperatures, the development of cyclones or other extreme weather conditions, and more (Emberson et al., 2018).

5.3 Nitrogen Oxide (NO_x)

Nitrogen oxide is a pollutant commonly associated with vehicles due to its emission from automobile engines. When inhaled, it can deeply penetrate into the lungs and act as an irritant to the respiratory system, leading to respiratory diseases, coughing, congestion, shortness of breath, bronchospasm and even pulmonary edema at high concentrations. Adverse effects in humans are often observed at concentrations exceeding 0.2 parts per million (ppm), while concentrations above 2.0 ppm can impact specific immune system cells such as CD8+ cells and NK cells. Chronic respiratory problems have been related to protracted exposure to high amounts of nitrogen dioxide (NO_2). The ability to smell may be harmed over time by NO_2 exposure. There have been occurrences of ocular, throat, and nasal irritation in addition to respiratory system discomfort. Nitrogen dioxide has been discovered to have negative impacts on vegetation and crops, lowering crop output and diminishing plant growth efficiency. Additionally, NO_2 can taint materials and reduce visibility (Chen et al., 2007).

5.4 *Sulphur Dioxide (SO_2)*

Sulphur dioxide is a toxic gas primarily emitted through the combustion of fossil fuels and industrial processes. The standard annual concentration of SO_2 is typically around 0.03 parts per million (ppm) (US EPA). Table of Historical SO_2 NAAQS, Sulfur US EPA. Available online at: https://www3.epa.gov/ttn/naaqs/standards/so2/s_so2_history.html (accessed June 5, 2023). It can have detrimental effects on humans, animals, and plants. Individuals who are more susceptible, such as those with pre-existing pulmonary conditions, the elderly and infants, are at a higher risk. In industrialized areas, the major health issues associated with sulphur dioxide emissions include respiratory tract irritation, bronchitis, increased mucus production and bronchospasm. Sulphur dioxide may profoundly permeate the lungs, where it is converted to bisulfite and interacts with sensory receptors to cause bronchoconstriction. Additionally, it can result in skin redness, eye damage (tears and corneal opacity), mucous membrane irritation, and the worsening of preexisting cardiovascular problems (Goshua, 2022). Environmental effects of sulphur dioxide emissions include soil acidification and the development of acid rain.

5.5 *Lead*

Lead is a heavy metal that is used in a number of industrial processes and released into the environment by a number of gasoline-powered engines, batteries, radiators, waste incinerators and wastewater. Additionally, sources including metals, mining, and piston-engine aircraft can contribute to lead pollution in the atmosphere. Because lead poisoning harms people, animals and the environment, it poses a serious risk to public health, especially in developing nations. Lead exposure can be by ingesting, inhalation or skin absorption. Trans-placental transportation can occur from it being able to cross the placenta. The harmful consequences of lead are particularly dangerous for children. Young children are more adversely affected by toxic effects. Lead poisoning damages the developing foetus' nervous system and causes swelling or edoema of the brain. Lead builds up in the blood, soft tissues, liver, lungs, bones, neurological system, cardiovascular system, and reproductive system after breathing. Adults may have symptoms including difficulty in concentrating, memory loss, and discomfort in their muscles and joints. Lead is a neurotoxic that, especially in children and newborns who are very vulnerable even to low levels (Farhat et al., 2013), can result in learning problems, memory impairment, hyperactivity, and even mental retardation. The growth of plants and crops is affected by the presence of high quantities of lead in the environment. Lead exposure in vertebrates and other animals at high levels is linked to neurological damage.

5.6 Polycyclic Aromatic Hydrocarbons (PAHs)

The main transport method for polycyclic aromatic hydrocarbons (PAHs), which are widely distributed in the environment, is atmospheric dispersion. They can be discovered in bituminous sediments and coal. In addition, incomplete combustion of organic material occurs in engine exhaust, incinerator operations, and forest fires, which results in the production of PAHs. Benzopyrene, acenaphthylene, anthracene, and fluoranthene are only a few PAHs that are known to be poisonous, mutagenic and carcinogenic. They pose a serious risk for developing lung cancer (Abdel-Shafy and Mansour, 2016).

5.7 Volatile Organic Compounds (VOCs)

An elevated risk of cancer has been linked to human exposure to volatile organic compounds (VOCs) such as toluene, benzene, ethylbenzene and xylene. Higher VOC concentrations have been found in a variety of interior situations as a result of the usage of new goods and materials. VOCs have the potential to pollute indoor air and have a negative impact on people's health. Exposure to VOCs can have a detrimental effect on human health, both immediately and over time. The smell in indoor air are frequently caused by VOCs. Long-term exposure to VOCs can have more harmful toxic consequences whereas short-term contact might irritate the eyes, sinuses, throat and mucous membranes. Given that these contaminants might interact in ways that are additive, antagonistic or neutral, it can be difficult to predict the harmful consequences of complex combinations of VOCs (Ebersviller et al., 2012).

5.8 Dioxins

Dioxins are produced by both industrial and natural activities, such as volcanic eruptions and forest fires. They have a propensity to build up in a variety of dietary sources, notably in the fatty tissues of animals. These sources include meat, dairy products, fish and shellfish. Skin lesions and discolouration can result from short-term exposure to high quantities of dioxins. On the other side, protracted exposure to dioxins may cause anomalies in the brain system, infertility, endocrine problems, immune system malfunction and endocrine disorders, and an elevated risk of cancer. There is no denying that burning fossil fuels contributes significantly to air pollution. Human activities including transportation, industrial and agricultural operations, and natural sources can all result in pollution (World Health Organization). Dioxins and their effects on human health. Available online at: https://www.who.int/news-room/fact-sheets/detail/dioxins-and-their-effects-on-human-health (accessed June 5, 2023).

6 Impact of Air Pollutants on Health

The most prevalent air pollutants are ozone at ground level and PM. Air pollutants can be widely categorized into two: (a) ambient air pollution, which refers to outdoor pollutants and (b) indoor pollution, which is due to the combustion of hydrocarbons inside houses. People exposed to excessive levels of air pollutants have various disease signs and conditions of various severity. Those health-associated results may be labeled as short- or long-time period effects. Certain populations, which includes the elderly, babies, and people with pre-current conditions which include diabetes and coronary heart or lung diseases (mainly bronchial asthma), are specifically susceptible and must take precautions to guard their fitness. But the precise magnitudes of both brief-time period and lengthy-time period influences have no longer been completely determined because of versions in epidemiological methodologies and potential exposure mistakes, as formerly highlighted. New models are being proposed to extra successfully examine on the spot and lengthy-time period human publicity records (Chen, 2007). This section defines the most common quick-time period and long-term health results, as well as general worries associated with both forms of outcomes, spotting that their incidence can often depend on factors including the environment, dosage and individual susceptibility.

Air pollution's short-term effects can vary from minor problems like irritation of eyes, nose, skin and throat to more serious illnesses including asthma, pneumonia, bronchitis and cardiovascular problems. Additionally, a brief exposure to air pollution can cause nausea, dizziness and headaches. Pollutant exposure over a long period can harm the nervous, reproductive and respiratory systems, increasing the risk of cancer and, in rare situations, fatality. These chronic long-term consequences can last for many years or even for lifetime. Additionally, chronic exposure to certain air contaminants may accelerate the growth of a number of cancers (Nakano and Otsuki, 2013). As mentioned before, inhalation of air pollutants is closely linked to respiratory disorders. These pollutants enter the body through the airways and can accumulate in cells, causing damage that is proportional to the specific contaminant and the dose involved. The health effects of air pollution can vary significantly depending on the country, region, season, and time of exposure. Long-term health impacts are more likely to occur in combination with prolonged exposure to pollutants and the aforementioned conditions. Individuals who have respiratory diseases like asthma are at a higher risk for injury from pollutants like PM, pollen, benzene and ozone (Guarnieri and Balmes, 2014). In general, those who already have health issues are more susceptible to the long-term impacts of air pollution. Acute tracheal irritation brought by polluted air can also cause voice change. Chronic obstructive pulmonary disease (COPD), which considerably raises morbidity and death rates, has been related to air pollution development. The main causes of COPD include prolonged exposure to industrial and traffic-related air pollution as well as fuel combustion (Jiang et al., 2016).

Numerous cardiovascular issues (Bourdrel et al., 2017) have been recorded after exposure to air pollution. Cardiovascular function may be hampered by blood cell alterations brought on by prolonged exposure. Long-term exposure to traffic emissions is associated to coronary arteriosclerosis, but short-term exposure is linked to hypertension, stroke, myocardial infarctions and cardiac insufficiency. Human ventricular hypertrophy has been linked to long-term nitrogen oxide (NO_2) exposure (Leary et al., 2014).

Adults and infants neurological outcomes from prolonged exposure to air pollution have been related. It is linked to mental health, autism, retinopathy, foetal development abnormalities and low birth weight. Although the precise origin of neurodegenerative disorders like Alzheimer's and Parkinson's is uncertain, persistent air pollution exposure is thought to be a significant factor. A few etiological variables include food, metals and pesticides. Inflammation, protein aggregation, oxidative stress and mitochondrial dysfunction in neurons are shown in Fig. 1 that contribute to the development of neurodegenerative diseases (Genc et al., 2012). Inflammation of the brain has been observed in long-term residents of heavily polluted regions in Mexico (Calderon-Garciduenas et al., 2002). Immediate response to ultrafine particulate matter includes elevated levels of systemic inflammation markers (IL-6 and fibrinogen) in adults, potentially leading to acute-phase protein production. Long-term air pollution is thought to cause neurological disturbances by promoting atherosclerosis progression and oxidative stress. Multiple organs' embryonic maturation is hampered by inflammation as a result of oxidative stress. Susceptibility to long-term air pollution depends on a number of variables, viz. birth weight, mother smoking, genetic background, socioeconomic situation and educational attainment. Another determining aspect is diet, starting with lactation. By scavenging free radicals and preventing their interaction in the brain, dietary antioxidants help to protect against air pollution. Susceptibility to oxidative stress is influenced by genetic background as well. In asthmatic children with a particular genetic genotype, antioxidant treatment with vitamins C and E can regulate the impact of ozone. The innate immune system's Toll-like receptor 2 is activated by inflammatory cytokines released in the periphery, such as respiratory epithelial. Rats exposed to ambient PM in Los Angeles, this pathway was activated, and subsequent events resulted in neurodegeneration (Rückerl et al., 2007). Lead exposure causes neurodevelopmental defects in children like aggressive behaviour, decreased intellect, learning challenges and hyperactivity. There is currently no established safe amount of lead exposure, and the scientific community has pushed the Centres for Disease Control and Prevention

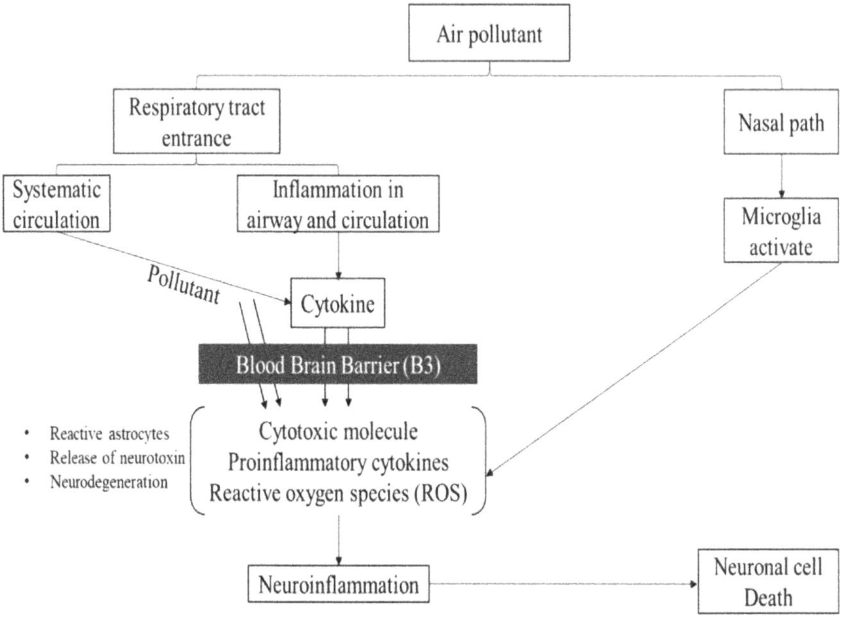

Figure 1: Brain damage caused by exposure to air pollution.

(CDC) to decrease the 10 g/dl screening threshold (Bellinger, 2008). The immune system is impacted by poor air quality, which causes dysregulation and neuro inflammation (Fig. 1). Nevertheless, elevated levels of complement component C_3 and immunoglobulins (IgA, IgM) are seen in the serum. Air pollution affects how an antigen is presented because macrophages overexpress costimulatory molecules including CD80 and CD86 (Nakano, 2013).

Pollution associated with traffic, which include PAHs, VOCs, oxides and PM, might also set-off pigmented patches on our skin. The organ damage is located while pollutants are absorbed via the epidermis or inhaled, as a number of those pollutants are mutagenic and carcinogenic and specifically affect the liver and lungs. In polluted city, air pollutants (and those inside the troposphere) lessen the dangerous consequences of ultraviolet radiation. Smoking and exposure to PM speed up the ageing process of the epidermis, resulting in wrinkles, dyschromia and spots. Finally, skin cancer has been connected to environmental contaminants (Drakaki et al., 2014). When exposed to the above hazards, foetuses and babies have higher risk of morbidity. Autism, prenatal development retardation, and low birth weight have all been recorded (Weisskopf et al., 2015). Another external organ that might be harmed by air pollution is the eye. Contamination from dispersed pollutants can lead to various eye effects, including asymptomatic conditions, irritation, retinopathy, or dry eye syndrome (Mo et al., 2019).

7 Discussion and Conclusion

Director-General Dr. Tedros Adhanom Ghebreyesus of WHO referred to air pollution as "the new tobacco" and described it as a "silent public health emergency" during the first WHO Global Conference on Air Pollution and Health in 2018 (available online at: https://www.who.int/airpollution/events/conference/en/ (accessed June 6, 2023)). Children are undoubted most susceptible to the impacts of air pollution in their formative years. Numerous detrimental effects of air pollution on our daily life is well documented. In addition to heavy financial burden on society, the illnesses brought by air pollution also have social repercussions including missed days at work and in school. Although addressing human-caused environmental pollution is challenging, a collaborative effort by authorities, organizations and healthcare professionals is also important. Governments should prioritize the dissemination of adequate information, public education and the involvement of experts to effectively prevent the emergence of air pollution problems. It is imperative to develop and implement technologies that reduce air pollution at its source in all industries and power facilities.

The Kyoto Protocol of 1997 set the important goal of bringing greenhouse gas emissions down to 5% below 1990 levels by 2012 (What is the Kyoto Protocol? UNFCCC. Available online at: https://unfccc.int/kyoto__protocol (accessed June 6, 2023)). International summits that followed, including 2009 in Copenhagen and 2011 in Durban, strengthened the commitment to this strategy. Both the Kyoto Protocol and subsequent accords have received widespread ratification. A recent significant international accord addressing climate change was the 2015 Paris accord, published by the UNFCCC (United Nations Framework Convention on Climate Change). This agreement has been ratified by numerous United Nations and European Union member states. Those involved in these agreements should advocate for actions and measures that improve various aspects related to climate change and environmental pollution. To increase the likelihood of attaining goals and objectives regarding this topic, it is imperative to increase education, training, public knowledge, and engagement (Paris Climate Change Agreement. (2016). Available online at: https://unfccc.int/process-and-meetings/the-paris-agreement/the-paris-agreement (accessed June 6, 2023)).

While it may be challenging to reduce the negative impact of greenhouse gas emissions, technological advancements provide opportunities to limit their use by adopting reliable methods. In addition to managing the adverse health effects of air pollution, it is important to develop a global prevention policy to combat anthropogenic air pollution. This requires employing sustainable development practices and utilizing research-based knowledge to effectively address the issue (Manisalidis et al., 2020).

Global collaboration in terms of research, development, management policy, monitoring, and politics is increasingly necessary for effective pollution control.

Policymakers should advocate for the creation of a potent instrument for environmental and health protection, and air pollution control should be harmonised and modernised.

Key Take Home Messages

1. Long-term and short-term exposure to PM is associated with negative health consequences. Compared to PM_{10} particles, $PM_{2.5}$ particles are a bigger hazard to human health.
2. Long-term exposure to PM is linked to immune system problems, neonatal mortality, respiratory illnesses, and cardiovascular ailments. People with respiratory and cardiovascular disorders, diabetes, asthma, pneumonia, and other diseases are especially susceptible to the negative effects of PM.
3. The health of people, plant development, and vegetation may be negatively impacted by ozone (O_3) levels in metropolitan areas. Plant development is slowed by ozone, which also causes biochemical, morphological, functional, and immunological issues and damages DNA and inhibits cellular function.
4. Air pollutants that harm both human health and the environment include lead, polycyclic aromatic hydrocarbons (PAHs), volatile organic compounds (VOCs), carbon monoxide (CO), nitrogen dioxide (NO_2), sulphur dioxide (SO_2), and dioxins.
5. Respiratory conditions including asthma and bronchitis, as well as eye, nose, and skin irritation, are some of the short-term health repercussions of air pollution. Chronic respiratory conditions, cardiovascular conditions, and neurological problems can result from prolonged exposure to air pollution. Particularly at danger are vulnerable groups like elderly, young children, and people with pre-existing medical issues.

References

Abdel-Shafy HI, Mansour MSM (2016) A review on polycyclic aromatic hydrocarbons: source, environmental impact, effect on human health and remediation. Egypt J Pet 25:107–123. https://doi.org/10.1016/j.ejpe.2015.03.011

Bellinger DC (2008) Very low lead exposures and children's neurodevelopment. Curr Opin Pediatr 20:172–177. https://doi.org/10.1097/MOP.0b013e3282f4f97b

Bezirtzoglou C, Dekas K, Charvalos E (2011) Climate changes, environment and infection: facts, scenarios and growing awareness from the public health community within Europe. Anaerobe 17:337–340. https://doi.org/10.1016/j.anaerobe.2011.05.016

Boschi N (ed) (2012) Defining an educational framework for indoor air sciences education. In: Education and training in indoor air sciences. Springer Science & Business Media, Luxembourg, 245p

Bourdrel T, Bind M-A, Béjot Y, Morel O, Argacha J-F (2017) Cardiovascular effects of air pollution. Arch Cardiovasc Dis 110:634–642. https://doi.org/10.1016/j.acvd.2017.05.003

Calba C, Guerbois-Galla M, Franke F, Jeannin C, Auzet-Caillaud M, Grard G, Pigaglio L, Decoppet A et al (2017) Preliminary report of an autochthonous chikungunya outbreak in France, July to September 2017. Eur Surveill 22:17–00647. https://doi.org/10.2807/1560-7917.ES.2017.22.39.17-00647

Calderon-Garciduenas L, Azzarelli B, Acuna H et al (2002) Air pollution and brain damage. Toxicol Pathol 30:373–389. https://doi.org/10.1080/01926230252929954

Chen T-M, Gokhale J, Shofer S, Kuschner WG (2007) Outdoor air pollution: nitrogen dioxide, sulfur dioxide, and carbon monoxide health effects. Am J Med Sci 333:249–256. https://doi.org/10.1097/MAJ.0b013e31803b900f

Drakaki E, Dessinioti C, Antoniou C (2014) Air pollution and the skin. Front Environ Sci Eng China 15:2–8. https://doi.org/10.3389/fenvs.2014.00011

Ebersviller S, Lichtveld K, Sexton KG, Zavala J, Lin Y-H, Jaspers I et al (2012) Gaseous VOCs rapidly modify particulate matter and its biological effects – part 1: simple VOCs and model PM. Atmos Chem Phys Discuss 12:5065–5105. https://doi.org/10.5194/acpd-12-5065-2012

Emberson LD, Pleijel H, Ainsworth EA, den Berg M, Ren W, Osborne S et al (2018) Ozone effects on crops and consideration in crop models. Eur J Agron 100:19–34. https://doi.org/10.1016/j.eja.2018.06.002

Ethical Unicorn (2019) Everything you need to know about aerosols & air pollution. https://ethicalunicorn.com/2019/04/29/everything-you-need-to-know-about-aerosols-air-pollution/. Accessed 4 Oct 2019

Farhat A, Mohammadzadeh A, Balali-Mood M, Aghajanpoor-Pasha M, Ravanshad Y (2013) Correlation of blood lead level in mothers and exclusively breastfed infants: a study on infants aged less than six months. Asia Pac J Med Toxicol 2:150–152

Genc S, Zadeoglulari Z, Fuss SH, Genc K (2012) The adverse effects of air pollution on the nervous system. J Toxicol 2012:782462. https://doi.org/10.1155/2012/782462

Goshua A, Akdis CA, Nadeau KC (2022) World Health Organization global air quality guideline recommendations: executive summary. Allergy 77(7):1955–1960. https://doi.org/10.1111/all.15224. Epub 2022 Feb 8 PMID: 35060140

Gryparis A, Forsberg B, Katsouyanni K, Analitis A, Touloumi G, Schwartz J et al (2004) Acute effects of ozone on mortality from the "air pollution and health: a European approach" project. Am J Respir Crit Care Med 170:1080–1087. https://doi.org/10.1164/rccm.200403-333OC

Guarnieri M, Balmes JR (2014) Outdoor air pollution and asthma. Lancet 383:1581–1592. https://doi.org/10.1016/S0140-6736(14)60617-6

Heal MR, Kumar P, Harrison RM (2012) Particles, air quality, policy and health. Chem Soc Rev 41:6606–6630. https://doi.org/10.1039/c2cs35076a

Jacobson MZ, Jacobson PMZ (2002) Atmospheric pollution: history, science, and regulation. Cambridge University Press, p 206. https://doi.org/10.1256/wea.243.02

Jiang X-Q, Mei X-D, Feng D (2016) Air pollution and chronic airway diseases: what should people know and do? J Thorac Dis 8:E31-40

Kloog I, Ridgway B, Koutrakis P, Coull BA, Schwartz JD (2013) Long- and short-term exposure to PM2.5 and mortality using novel exposure models. Epidemiology 24:555–561. https://doi.org/10.1097/EDE.0b013e318294bcaa

Leary PJ, Kaufman JD, Barr RG, Bluemke DA, Curl CL, Hough CL et al (2014) Traffic-related air pollution and the right ventricle. The multi-ethnic study of atherosclerosis. Am J Respir Crit Care Med 189:1093–1100. https://doi.org/10.1164/rccm.201312-2298OC

Manisalidis I, Stavropoulou E, Stavropoulos A, Bezirtzoglou E (2020) Environmental and health impacts of air pollution: a review. Front Public Health 8:14. https://doi.org/10.3389/fpubh.2020.00014. PMID:32154200; PMCID: PMC7044178

Mo Z, Fu Q, Lyu D, Zhang L, Qin Z, Tang Q et al (2019) Impacts of air pollution on dry eye disease among residents in Hangzhou, China: a case-crossover study. Environ Pollut 246:183–189. https://doi.org/10.1016/j.envpol.2018.11.109

Moller L, Schuetzle D, Autrup H (1994) Future research needs associated with the assessment of potential human health risks from exposure to toxic ambient air pollutants. Environ Health Perspect 102(Suppl. 4):193–210. https://doi.org/10.1289/ehp.94102s4193

Nakano T, Otsuki T (2013) Environmental air pollutants and the risk of cancer. (Japanese). Gan to Kagaku Ryoho 40:1441–1445

Paris Climate Change Agreement (2016). https://unfccc.int/process-and-meetings/the-paris-agreement/the-paris-agreement. Accessed 6 June 2023 [Ref list]

Rückerl R, Greven S, Ljungman P, Aalto P, Antoniades C, Bellander T et al (2007) Air pollution and inflammation (interleukin-6, C-reactive protein, fibrinogen) in myocardial infarction survivors. Environ Health Perspect 115:1072–1080. https://doi.org/10.1289/ehp.10021

Stansfeld SA (2015) Noise effects on health in the context of air pollution exposure. Int J Environ Res Public Health 12:12735–12760. https://doi.org/10.3390/ijerph121012735

US EPA. Table of Historical SO_2 NAAQS, Sulfur US EPA. https://www3.epa.gov/ttn/naaqs/standards/so2/s_so2_history.html. Accessed 6 June 2023 [Ref list]

Weisskopf MG, Kioumourtzoglou M-A, Roberts AL (2015) Air pollution and autism spectrum disorders: causal or confounded? Curr Environ Health Rep 2:430–439. https://doi.org/10.1007/s40572-015-0073-9

What is the Kyoto Protocol? UNFCCC. https://unfccc.int/kyoto__protocol. Accessed 6 June 2023 [Ref list]

WHO (World Health Organization). Dioxins and their effects on human health. https://www.who.int/news-room/fact-sheets/detail/dioxins-and-their-effects-on-human-health. Accessed 6 June 2023 [Ref list]

WHO. First WHO Global Conference on Air Pollution and Health (2018). https://www.who.int/airpollution/events/conference/en/. Accessed 6 June 2023 [Ref list]

Wilson WE, Suh HH (1997) Fine particles and coarse particles: concentration relationships relevant to epidemiologic studies. J Air Waste Manag Assoc 47:1238–1249. https://doi.org/10.1080/10473289.1997.10464074

Isotope Fingerprinting in Air Pollution Source Identification: A Review

Mobarok Hossain, Parul Kuiry, and P. K. Patra

1 Introduction

The alteration of the atmosphere concerning air pollution is predominantly attributed to human- activities, including industrial and agricultural practices. The dynamics of precipitation and aerosol particles also influence the fluctuations in ground-level air pollutants. The assessment and forecasting of concentration and dispersion of contaminant relies on our understanding of the underlying processes. Applying isotopes (radioactive and stable) as tracers of atmospheric processes, including source identification of air pollutants, is an emerging technology. The utility of isotopes is based on isotopic fractionation, i.e. the unequal distribution of light and heavy isotopes among the reactants and products of a chemical reaction. Isotopic fractionation can manifest through reversible equilibrium or irreversible unidirectional kinetic reactions (Kendall, 1998). Hence, the isotopic composition of a specific rainwater or aerosol not only mirrors the composition of the initial source or a combination of sources with distinct compositions but is also affected by the isotopic fractionation during the transportation and chemical conversion of the compounds (Nestler et al., 2011).

Isotopic measurement in aerosol and rainwater helps understand the dynamics of contaminant dispersion across the atmosphere, hydrosphere and lithosphere. The radioactive isotopes tritium and carbon-14 and the stable isotopes such as deuterium, oxygen, nitrogen and chlorine are commonly used in atmospheric research. While radioactive isotopes provide clues to residence time, stable isotopes are valuable

M. Hossain
Department of Applied Geosciences, University of Göttingen, Göttingen, Germany

P. Kuiry · P. K. Patra (✉)
Department of Environmental Studies, Visva-Bharati, Santiniketan, West Bengal, India
e-mail: pulakpatra@visva-bharati.ac.in

P. K. Padhy et al. (eds.), *Airborne Particulate Matter*,
https://doi.org/10.1007/978-3-031-84408-9_3

indicators for water circulation. Air contaminants with varying isotopic compositions frequently originate from different locations and times. Besides, environmental isotopes rarely interact significantly with other pollutants. Thus, isotopic variations typically exhibit identifiable and predictable patterns, facilitating their reconstruction.

This article provides a brief synopsis of the background information on using isotope tracers in air pollution. This study underscores the significance of the interconnectedness between the hydrosphere, atmosphere and lithosphere in shaping the isotopic composition, hence enabling the identification of many variables that contribute to air pollution. The article also deals with analytical procedures, calculations and interpretation of isotopic data to identify pollution sources.

2 Isotope in Rainwater

Atmosphere plays a significant role in the hydrological cycle by providing a necessary condition for water vapour condensation, leading to cloud formation and precipitation in either liquid or solid form. Atmospheric aerosols are the primary contributors to the chemical elements in rainfall. The aerosols include crustal and volcanic dust, sea salts, biogenic materials, and anthropogenic emissions. Rainwater chemistry is significantly influenced by atmospheric chemistry (Negrel et al., 2007).

Negrel et al. (2007) asserted an interaction between different aerosol types and atmospheric gases in the context of rainwater chemistry, suggesting rainwater's potential to serve as a viable source of trace elements. Numerous scholars are working towards the field of precipitation chemistry. Their findings indicate that rainwater tends to exhibit acidity predominantly because of the substantial presence of sulphate and nitrate aerosols, which can be traced back to human activities associated with the rapid expansion of the economy and the escalating consumption of fossil fuels (Rao et al., 2015). Identifying contamination sources and exploring their chemistry in a wide range sometimes involves isotopic analysis of sulphate and nitrate, both of anthropogenic and natural origin (Fenech et al., 2012).

Isotopic analysis of rainfall enables the differentiation of moisture derived from land and ocean because of discrete isotopic signatures. Besides, isotope information in the rain and air vapour can be employed to replicate the isotopic composition of moisture sources that have undergone fractionation during ocean-to-continental transit (Dansgaard, 1964; Araguas-Aragua et al., 2000; Yoshimura, 2015).

The presence of anthropogenic particles, such as heavy metals found in aerosols, might lead to the dissolution of these particles under low pH conditions, hence contributing to the contamination of rainfall. The water cycle encompasses various stages, including evaporation, transit, condensation, mixing, and distinct isotopic reservoirs, which can be used to identify a specific signature (Nlend et al., 2020).

The initial source of water vapour and the processes of fractionation that occur during droplet formation and subsequent descent through air masses are crucial factors in determining the precipitation of δw (the isotopic composition of water).

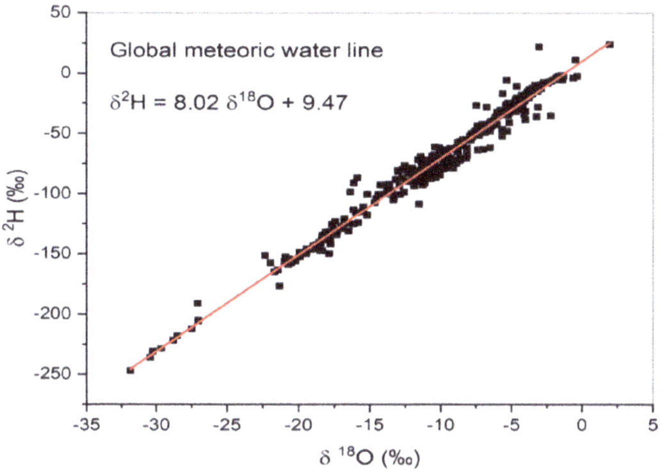

Figure 1: Global meteoric water line (MWL) (Craig, 1961).

Fractionation, which refers to alterations in the relative abundance of heavy and light isotopes, occurs during phase transitions and transportation due to the differential diffusion and evaporation of lighter water molecules compared to heavier ones and the preferential condensation of heavier water molecules. Fractionation occurs in equilibrium conditions, specifically at a constant dependent on temperature, in saturated environments characterised by 100 per cent relative humidity (RH). This phenomenon is commonly observed during the creation of rain droplets, where the evaporation and condensation rate are equal and is described through the worldwide annual average (global meteoric water line) ratio between hydrogen and oxygen isotopes (Oxygen-18 and Deuterium) in naturally occurring meteoric water (Fig. 1).

The meteoric water line (MWL) is defined by

$$\delta^2H = \left(8.02 \times \delta^{18}O\right) + d$$

where δ is the disparity in the ratio of isotope abundances ($^{18}O/^{16}O$ or $^2H/^1H$) compared to a reference isotope abundance ratio ($^{18}O/^{16}O$ or $^2H/^1H$), such as the Viena Standard Mean Ocean Water (V-SMOW).

The analysis of element concentration and pH levels in rainwater can yield valuable insights into the dissolved chemical characteristics present in the precipitation. The isotopic ratio of Ca, Mg and Sr offer greater utility in identifying the sources of aerosol fingerprints. This is because the metal ions and pH levels are influenced by aerosols' chemical composition and concentration, their solubility in rainwater, and the presence of atmospheric gases such as SO_2 and CO_2. The $^{87}Sr/^{86}Sr$ ratio is employed as an indirect method to determine Ca ions' origins, vital constituents found in rainfall (Han et al., 2019).

3 Isotope in Aerosol

Aerosols in ambient air can negatively affect living organisms (Kim et al., 2015). Aerosols originate from both anthropogenic and natural activities. The primary sources of anthropogenic emissions include activities related to road traffic (exhaust emissions, wearing of tyres and brakes, road dust), industry (power plants, oil refineries, chemical industries), domestic heating, mining and construction work (Calvo et al., 2013) in urban areas. Farming and biomass burning emissions are the primary anthropogenic sources in rural areas. Tobacco smoking is also a significant source of aerosol. Natural sources of aerosols include emissions from soil, wildfire, volcanic eruptions, oceans, and plants. Each aerosol source is characterised by unique chemical and isotopic composition (Calvo et al., 2013). The chemical composition of aerosol particles influences their impact on atmospheric behaviour (Heintzenberg, 2003). Thus, the knowledge of composition, properties, sources and formation mechanisms of aerosols helps to understand their actual impact on human and ecosystem health (National Research Council, 2004; IPCC, 2007). Isotope analysis method have demonstrated its ability to identify the relative contribution of major three sources to atmospheric aerosols, i.e. fossil and non-fossil fuel sources from land areas and marine sources (Ceburnis et al., 2011).

4 Isotopic Fractionation and Interpretation

Accurately quantifying the abundance of isotopes is challenging due to the low occurrence of heavy isotopes. Consequently, the isotope ratio is ascertained as the proportion of heavier to lighter isotopes. "Isotopic abundance" pertains to the proportion or percentage of a specific isotope relative to other isotopes of the same element within a given system. The stable isotope composition of an element, denoted as delta (δ), is represented by the isotopic abundance ratio stated in parts per mil notation (‰) concerning the corresponding ratio in a globally recognised reference standard material (Vienna Pee Dee Belemnite) (Gröning, 2004). Various fractions of isotopes are employed to trace distinct origins of pollutants. Isotopic fractionation refers to the phenomenon when there is a significant disparity in the ratio of heavy isotopes to light isotopes during the reaction between a substance and its resulting product. The utilisation of various isotopic compositions identifies the origins of contaminants, including both the primary source and any subsequent mixing that may have occurred.

Additionally, it is essential to acknowledge that the chemical composition of compounds can influence isotopic fractionation as they undergo transportation and transformation processes (Fenech et al., 2012). The isotopic fractionation (ε) is mathematically defined as the difference between the isotopic fractionation factor α and unity. Typically, this quantity is of a diminutive magnitude; it is multiplied by a factor of 10^3 and, afterwards, denoted in parts per thousand (‰). The temperature-dependent nature of isotopic fractionation was first observed by Urey (1948), who observed a reduction in fractionation as temperature increased.

The isotopic distribution is also influenced by the type of chemical bonds. High ionic potential and low atomic masses show elevated vibrational frequencies and facilitate the preferential incorporation of heavier isotopes. For example, the calcite formed during inorganic precipitation consistently exhibits a higher concentration of the oxygen isotope ^{18}O compared to other minerals. This disparity can be attributed to the tiny, highly charged C^{4+} ion strongly attached to the oxygen. The crystal structure of a material also influences the preferential accumulation of heavier isotopes, with more ordered and densely packed structures exhibiting a greater tendency for such accumulation. The phenomenon of increased ^{18}O concentration in ice relative to water has been documented by Beck and Münnich (1988).

Two distinct categories of isotopic fractionation processes exist, specifically "equilibrium" and "kinetic."

Equilibrium Fractionation

This is a specific instance of a chemical equilibrium reaction when no overall reaction occurs, but rather an exchange of isotopes transpires, influencing the equilibrium constant. Thus, isotopes undergo movement, establishing equilibrium once the isotopic ratios cease exhibiting further changes over time. As an illustration, foraminifers, single-celled microorganisms, produce their calcite shells in equilibrium with the surrounding saltwater. During this process, an isotope exchange occurs.

$$1/3\mathrm{CaC}^{16}\mathrm{O}_3 + \mathrm{H}_2^{18}\mathrm{O} \rightleftharpoons 1/3\mathrm{CaC}^{18}\mathrm{O}_3 + \mathrm{H}_2^{16}\mathrm{O}$$

The equilibrium constant K represents the quantitative measure of the equilibrium position for a given reaction and is expressed as:

$$K = \frac{[\mathrm{CaC}^{18}\mathrm{O}_3][\mathrm{H}_2^{16}\mathrm{O}]}{[\mathrm{CaC}^{16}\mathrm{O}_3][\mathrm{H}_2^{18}\mathrm{O}]}$$

or

$$K = \frac{[(\mathrm{CaC}^{18}\mathrm{O}_3)/(\mathrm{CaC}^{16}\mathrm{O}_3)]^{1/3}}{[(\mathrm{H}_2^{18}\mathrm{O}/(\mathrm{H}_2^{16}\mathrm{O})]}$$

The equilibrium constant is related to the fractionation factor by $\alpha = K^{1/n}$, where n is the number of atoms that have been transferred.

Kinetic Fractionation

This phenomenon is linked to unidirectional and incomplete processes. Isotopic equilibrium is not achieved in instances of kinetic fractionation. In the context of a chemical reaction, the rate constants kL and kH represent distinct values associated with the lighter and heavier isotopes, respectively. The impact of isotopic substitution in the reactant can be quantified by the ratio of these rate constants, denoted as kL/kH. If the ratio exceeds 1, it indicates that the reactant with the lighter isotope has a higher reactivity and is referred to as "normal". The term "inverse" describes a situation where the ratio is smaller than 1, indicating that the reactant with a heavier isotope undergoes a faster reaction. In the context of an irreversible reaction, it can be inferred that the absence of any remaining reactants upon forming a product would preclude the occurrence of isotope effects and fractionation.

Isotope ratio mass spectrometers (IRMS) are utilised for isotopic analysis, employing the electromagnetic field-induced separation of charged molecules and atoms depending on their respective masses. The IRMS technique cannot accurately determine a substance's absolute isotopic mass with the required level of precision. However, it can accurately and precisely determine the disparity in isotopic ratios between two compounds, as they undergo identical conditions during the study. Hence, the isotopic abundances are expressed in the form of δ values, representing the relative deviation of the isotopic ratio of a given sample from that of an internationally recognised standard:

$$^{\delta}A = \left(\frac{R_A}{R_B}\right) - 1$$

The variables R_A and R_B represent the ratios of the less abundant (heavier) isotope to the more abundant (lighter) isotope in the sample and standard, respectively. For example, R_A may be the ratio of ^{18}O to ^{16}O in the sample, while R_B could be the ratio of ^{18}O to ^{16}O in the standard. The δ-value is a dimensionless quantity. It is multiplied by 10^3 and illustrated in "per mil" (‰) units for the simplicity of understanding. When representing isotopic ratios, placing the heavier isotope in the numerator is customary to ensure that the resulting δ-value is positive. A positive value for δ indicates that the sample contains a higher proportion of isotopes with greater mass than the standard.

For a substance to be suitable for utilisation as a reference or standard material, it must exhibit enough stability, homogeneity and widespread availability, and possess an isotope ratio that closely approximates the average value within the range of natural fluctuation. A substance created within a specific laboratory is called "lab standard." However, it is necessary to calibrate this substance against an international standard to ensure consistency and comparability. This calibration allows for reporting δ-values of samples from various locations worldwide on a standardised scale. The *primary reference materials* (international standards) for several light elements include Vienna Standard Mean Ocean Water (VSMOW) for isotopic ratios of oxygen-18 ($\delta^{18}O$) and deuterium ($\delta^{2}H$), Vienna Pee Dee Belemnite (VPDB) for isotopic ratios of oxygen-18 ($\delta^{18}O$) and carbon-13 ($\delta^{13}C$), air-N_2 for isotopic ratios of nitrogen-15 ($\delta^{15}N$), Vienna Canyon Diablo Troilite (an iron meteorite) for isotopic ratios of sulfur-34 ($\delta^{34}S$), SRM 951 (Boric Acid; formerly known as NBS-951) for isotopic ratios of boron-11 ($\delta^{11}B$), and LSVEC (a lithium carbonate) for isotopic ratios of lithium-7 ($\delta^{7}Li$), among other materials. Renowned organisations such

Table 1: Natural isotopic abundances and international reference standards

Element	*Stable isotopes*	*Abundance (%)*	*Sample gas*	*Reference standard*	*Value of standard*
Hydrogen	^{1}H, ^{2}H	99.985 0.015	H_2	Water (VSMOW)	$^{2}H/^{1}H =$ 155.76±0.1ppm
Bromine	^{79}Br, ^{81}Br	50.7 40.3	CH_3Br	Bromide (SMOB)	
Carbon	^{12}C, ^{13}C	98.89 1.11	CO_2	Carbonate (VPDB)	0.0112372
Chlorine	^{35}Cl, ^{37}Cl	75.78 24.22	CH_3Cl	Chloride (SMOC)	0.324
Nitrogen	^{14}N, ^{15}N	99.63 0.37	N_2	Air (AIR)	
Oxygen	^{16}O, ^{17}O, ^{18}O	99.759 0.037 0.204	CO	Water (VSMOW)	
Sulphur	^{32}S, ^{33}S, ^{34}S, ^{36}S	95.01 0.75 4.215 0.017	SO_2	Troilite (VCDT)	

VSMOW - Vienna Standard Mean Ocean Water.
SMOB - Standard Mean Ocean Bromide.
VPDB - Vienna Pee Dee Belemnite.
VCDT - Vienna Canyon Diablo Troilite.

as National Institute of Standards and Technology (NIST) in USA, International Atomic Energy Agency (IAEA) in Vienna, and Institute for Reference Materials and Measurements in Belgium are responsible for developing and disseminating international standards. These standards include "certified isotopic reference materials" such as VSMOW2, NBS 19 and IAEA N 1, which possess predetermined δ values. These materials are utilised for calibration to the international standard scale. Some isotopes standard value and their abundance are given in Table 1.

5 Instruments and Methods for Isotope Analysis

Historically, conventional methods for identifying and allocating sources have mostly revolved around utilising concentration gradients, and applying multivariate statistical analysis. Although they have shown promise thus far, they are prone to significant ambiguity in situations where there are a large number of emission sources, intricate environmental pathways and various physical and biogeochemical factors that eliminate any source indicators or modify their original composition.

The utilisation of isotopic analysis can discern a distinctive "fingerprint" that unveils the prevailing origin of the molecule. Isotopic signatures have practical applications in differentiating between anthropogenic and natural origins of a specific chemical, determining whether the compound is undergoing natural dissipation or degradation, identifying the remediation process involved (such as evaporation, oxidation, reduction, degradation), and assessing the mechanism, rate and extent of degradation (Schmidt et al., 2004).

The concept of isotope measurement is a dynamic field within analytical techniques. It involves the determination of isotopic composition for specific compounds through a sequence of processes, including extraction, separation, and purification of the target compounds. Multiple analytical techniques exist for quantifying the isotopic ratios of various chemicals. These techniques include inductively coupled plasma mass spectrometry (ICPMS), thermal ionisation mass spectrometry (TIMS) and the isotope ratio mass spectrometer (IRMS).

Until now, GC-IRMS, which involves the coupling of an IRMS with a gas chromatograph (GC) has been utilised in a range of investigations, often focusing on the stable isotope signatures of carbon ($^{13}C/^{12}C$ or $\delta^{13}C$) and hydrogen ($^{2}H/^{1}H$ or δD or $\delta^{2}H$). Like GC-MS, the separation of compounds in IRMS is achieved by exploiting the differences in mass. However, in IRMS, the separation is based on the subtle variations in mass between isotopes of the same substance. After preparing the analyte stream, it undergoes ionisation by an electron impact ion generator. Subsequently, the ionised analyte is shaped into an ion beam and propelled to a magnetic field. Within this field, the isotopes are segregated based on their respective mass-to-charge ratios. The ion beams that have been separated are subsequently detected by the utilisation of specialised Faraday collectors, with each collector dedicated to a specific ion beam. The resulting ion flow is amplified and transformed into a digital signal to facilitate further processing (Cincinelli et al., 2012).

6 Application of Isotopes in Air Pollution Investigation

6.1 Hydrogen

Hydrogen is characterised by the presence of two stable isotopes, ^{1}H and ^{2}H, commonly referred to as protium and deuterium. Utilising the hydrogen isotope proportion ($\delta^{2}H$ or δD) of volatile organic compounds (VOCs) in the atmosphere can serve as a valuable approach to identifying the primary source of these molecules. The variability in the δD ratio of n-alkanes in fossil fuels is significantly greater than that of their stable carbon isotope ($\delta^{13}C$), which can be attributed to differences in hydrogen sources and post-depositional activities. Consequently, utilizing δD proportions of n-alkanes in aerosols is a valuable tool for obtaining supplementary information regarding their origins and source patterns. This suggests that the δD values of n-alkanes can be employed in assessing the collective contribution of higher plants to fossil and fuel sources (Yamamoto & Kawamura, 2011).

Stable hydrogen isotope proportions (δ^2H) have been identified as effective tracers for volatile organic compounds (VOCs) (Rudolph et al., 2002). The mass differences between hydrogen isotopes are more pronounced than carbon isotopes. Higher isotopic contrasts allow disclosure of discrepancies across sources and photochemical responses in air. In contrast to the extensive research conducted on δ^{13}C values in volatile organic compounds (VOCs), a limited number of studies have investigated δ^2H values (Goldstein and Shaw, 2003). Several studies (Bilke and Mosandl, 2002; Jia et al., 2008) have conducted exact investigations on δ^2H values. Previous studies have examined the analysis of hydrogen isotope fractionation in several research areas, including biodegradation (USEPA, 2008), photochemical reactions (Poulson and Drever, 1999), and vaporisation (Wang and Huang, 2001, 2003). The estimation of δ^2H values for methane (Rice et al., 2001; Yamada et al., 2003), formaldehyde (Rice and Quay, 2006), and n-alkanes found in unrefined oils (Lu et al., 2005) and plants (Liu et al., 2006) has been conducted concerning ecological samples. Vitzthum et al. (2011a) estimated the δ^2H ratio of volatile organic compounds (VOCs) emitted by biomass ignition and petroleum and diesel vehicle exhaust. The findings revealed a significant disparity in the δ^2H values of VOCs originating from these different sources. The same research group estimated the δ^2H values of volatile organic compounds (VOCs) in the emissions from an alumina industry stack (Vitzthum et al., 2011b). It was observed that the δ^2H value of VOCs in the alumina stack emissions was higher compared to natural sources such as raw petroleum and sediments.

6.2 Carbon

Carbon in atmospheric aerosols is a significant factor in air pollution, posing potential risks to living creatures and the natural environment (Kampa and Castanas, 2008). Carbon isotopes have been employed in research on air pollution to define emissions originating from biomass burning and other fuel sources since 1980s (Court et al., 1981). Carbon has three isotopic variants that occur naturally: two stable isotopes (^{12}C and ^{13}C) and one radioactive isotope (^{14}C). In the natural environment, ^{12}C comprises 98.9% of the total carbon concentration, whereas ^{13}C and ^{14}C include 1.1% and 1.2×10^{-10} % (Ehleringer and Rundel, 1989). Understanding the contributions, emission sources, and atmospheric transformations of various carbon-containing aerosols remains limited. The analysis of multiple aerosols and their respective carbon percentages can yield valuable insights into the origins and mechanisms of aerosol generation.

The particulate matter (PM) composition encompasses various chemical constituents, among which total carbon (TC) is a major. TC is categorised into organic carbon (OC) and elemental carbon (EC). One significant contributor to the presence of EC particles is the emission of soot from the incomplete combustion of various sources, including biomass, wood and fossil fuels such as coal, oil and diesel. The chemical alteration of EC is minimal, making it a dependable biomarker of anthropogenic air contaminants (Ho et al., 2006). The composition of OC primarily consists of biogenic sources, including spores, pollen, bacteria and volatile compounds.

For formulating efficient measures for mitigating aerosol particle pollution, it is imperative to differentiate between fossil fuel combustion sources and biomass burning (Garbarienė et al., 2016). The $^{13}C/^{12}C$ carbon isotope ratio enables the discrimination between soil organic matter and fossil fuel emissions, hence facilitating researchers' more precise determination of the origin of air pollutants. Additionally, the $^{13}C/^{12}C$ isotope ratio has been employed to differentiate road traffic and industrial particle sources, as shown by Widory et al. (2004).

The isotopic composition $\delta^{13}C$in vehicles associated with traffic emissions falls within the range of -28.3 to -24.5‰ (Aguilera and Whigham, 2018). This range is comparable to non-vehicular fuel sources, including coal, charcoal, and fireplace emissions, which exhibit a $\delta^{13}C$ range of -30.1 to -23.4‰. With dust particles, the observed values varied from −21.0 to −16.4‰ for tiny particles ($PM_{2.5}$), whereas the range was narrower for coarse particles (PM_{10}), ranging from −26.9 to −25.1‰. This more limited range in PM_{10} may be attributed to the presence of organic material in the PM_{10} sample. Regarding biogenic sources, it was observed that C_3 plants exhibited a $\delta^{13}C$ range of -34.7 to -25.4‰, which was found to be the least enriched among the tested sources. In contrast, C_4 plants ranged from -22.2 to -13.0‰, indicating a substantially higher enrichment. The $\delta^{13}C$ range of mixed sources had values that closely approximated the average of each source.

It is anticipated that the carbon present in particulate matter (PM) resulting from vehicular traffic will exhibit an $\delta^{13}C$ value that closely aligns with the carbon isotope composition of the corresponding fuel utilised. The average $\delta^{13}C$ readings for diesel and gasoline were –27.6‰ and -26.2‰, respectively. In metropolitan settings, the isotopic composition of carbon-13 ($\delta^{13}C$) measures the collective contribution from several potential local sources, encompassing anthropogenic activities such as coal, petrol, and natural gas combustion. A higher $\delta^{13}C$ value is indicative of increased contributions from coal burning. In contrast, lower carbon concentrations may be attributed to a more significant prevalence of diesel-based road traffic than petrol-based traffic (Sharma et al., 2015).

One prevalent obstacle encountered in studies that employ $\delta^{13}C$ values to evaluate the origins of carbon in aerosols is overlapping ranges of $\delta^{13}C$ values observed in emissions from several sources, such as automobile and industry-related emissions. To address this issue, numerous scholars have employed integrated approaches to measuring. As an illustration, a particular study used $\delta^{13}C$ readings with air mass movement patterns to delineate and classify sources. In another investigation, both elemental compositions and carbon concentrations of pollutants were considered to measure these contaminants' diverse contributions. In a separate investigation, researchers employed the isotopic composition of oxygen ($\delta^{17}O$, $\delta^{18}O$) derived from samples of CO_2 to discern various origins of pollution and to delineate the nature of pollution incidents. The utilisation of $\delta^{18}O$ changes proved to be an efficient method for determining the characteristics of the source (fossil fuel or biomass combustion) and its geographical origin (local or regional/continental). Moreover, a study was conducted utilising lead isotope ratios to differentiate between sources of contamination from industrial activities and road traffic.

Using radioactive carbon (^{14}C) measurements and the $^{14}C/^{12}C$ isotope ratio facilitates the differentiation between the emissions from fossil fuel combustion and those from non-fossil sources. According to Hildemann et al. (1994), emissions originating from fossil fuels have a negligible presence of ^{14}C compared to contemporary sources. Radiocarbon (^{14}C) measurement distinguishes between biomass burning and biogenic emissions sources of carbonaceous aerosol. Most radiocarbon tests have been conducted on carbonaceous aerosol samples with particle sizes of $PM_{2.5}$ and PM_{10}, as documented by Dusek et al. (2013). One problem associated with radiocarbon analysis is its inability to differentiate between coal-derived aerosol particles and emissions from diesel and gasoline sources.

6.3 Nitrogen

Nitrogen exists in the atmosphere primarily as a substantial reservoir. However, in air particles and aerosols, nitrogen is also found in the form of NO_x, obtained from many sources such as coal-fired facilities, residential coal burning, and automobile emissions. On the other hand, within the natural ecosystem, the formation of NO_x can occur through many processes, such as biomass burning (Andreae and Merlet, 2001; lightning (Schumann and Huntrieser, 2007), Delmas et al., 1997), and soil processing (Williams et al., 1992). The anthropogenic emissions have led to a significant increase in the atmospheric concentration of nitrogen oxides (NO_x), primarily composed of NO and NO_2, over the past few decades (Galloway et al., 2004, 2008).

The atmospheric nitrogen oxide (NO_x) cycle is a multifaceted process that includes several mechanisms (Calvert et al., 1985; Jacob, 1999) that finally yield NO_3^- as the final product (Guha et al., 2017). Nitrate aerosol is generated in the atmosphere via chemical processes using ammonia (NH_3) and nitric acid (HNO_3) as the primary ingredients. The formation of nitric acid can occur through the reaction of NH_3 and nitrogen oxides (NO_x) derived from many sources (Guha et al., 2017). The level of NO_x concentration directly impacts the rate at which ozone is formed and the creation of particulate matter in the atmosphere. Therefore, it has significant ramifications for air quality (Ehhalt et al., 2001). After release into the atmosphere, NO_x molecules may undergo a series of oxidation reactions, ultimately transforming into nitric acid. This conversion facilitates their removal through several deposition mechanisms (Ehhalt et al., 2001; Elliott et al., 2009). The transformation of nitrogen oxides (NO_x) into aerosol nitrate renders it a viable subject for scientific investigations aimed at comprehending the sources of NO_x and its processes of generation and depletion throughout time. Additionally, the transportation of aerosol nitrate over extended distances might result in its deposition in nutrient-depleted regions, such as the Pacific Ocean, exerting an influence on primary production (Hsu et al., 2012).

Extensive research has been conducted on the isotopic composition of nitrate in several natural settings, including the atmosphere, hydrosphere, and geosphere.

Using isotopic measurements to determine aerosols' total nitrogen, nitrate, and ammonium levels has proven valuable in discerning their origins and the various physical and chemical processes they undergo within the nitrogen cycle. This approach has been explored in several studies, including those conducted by Elliott et al. (2009), Felix et al. (2012), Heaton (1987), Mkoma et al. (2014), Pavuluri et al. (2010), Savarino et al. (2007), and Yeatman et al. (2001a, b). Analysing nitrogen and oxygen isotope ratios in nitrate ($\delta^{15}N$, $\delta^{18}O$, and $\Delta^{17}O$) allows insights into the environment's reaction processes, fluxes, sources, and nitrate transport.

Prior research has indicated that the $\delta^{15}N$-NO_x values linked to coal combustion varied between 9‰ and 26‰, as Felix et al. (2012) and Heaton (1990) reported. According to previous studies conducted by Walters et al. (2015) and Walters and Michalski (2015), the nitrogen isotope composition ($\delta^{15}N$ value) of nitrogen oxides (NO_x) emitted by automobiles and diesel-powered vehicles has been seen to vary between the ranges of -13‰ to 9‰ and -21‰ to -2‰, respectively. The range of $\delta^{15}N$ values associated with biomass burning has been shown to span from -7‰ to 12‰, according to Fibiger and Hastings (2016). In contrast, the $\delta^{15}N$ values attributed to microbial nitrogen cycling processes such as nitrification and denitrification have been reported to fall within the range of -60‰ to -20‰ by Li and Wang (2008), Yu and Elliott (2017), and Jin et al. (2021).

6.4 Oxygen

The stable oxygen isotopes ^{16}O, ^{17}O, and ^{18}O serve as tracers in the biosphere, allowing for tracking various oxygen-containing compounds. Their composition in atmospheric nitrate is a powerful indicator for analysing oxidation mechanisms. ^{18}O enrichment and excess ^{17}O have been detected in air nitrate samples worldwide. Efforts to simulate the enrichment of oxygen isotopes by utilising mass balance equations and photochemical models have resulted in satisfactory concordance between observed data and computational simulations. Hence, the utilisation of oxygen isotopes in atmospheric nitrate can be employed to evaluate alterations in atmospheric chemistry, serving as tracers in research on nitrate deposition.

Numerous research investigations have examined the regional and seasonal fluctuations of oxygen isotopes within atmospheric nitrate. The initial investigation conducted by Durka et al. (1994) was to assess $\delta^{18}O$ levels in nitrate present in precipitation. The study revealed atmospheric nitrate $\delta^{18}O$ values were higher than V-SMOW values, ranging from +55 to 72‰. Noteworthy fluctuations were noted across different seasons, with raised $\delta^{18}O$ values during colder months and lower values during warmer months. At high latitudes, there is a noticeable increase in the $\delta^{18}O$ levels seen in atmospheric nitrates compared to those observed at mid-latitudes. The initial measurement of $\delta^{17}O$ in atmospheric nitrate was documented by Michalski et al. (2003) based on the analysis of aerosols collected over one year in California. The elevated levels seen during the winter season were ascribed to the chemical process of ozone oxidation, which occurs during the conversion of nitrate to nitric acid. The polar regions exhibited elevated levels in comparison to the mid-latitudes. The fascinating aspects lie in the seasonal and regional patterns observed in the values of air nitrate $\delta^{17}O$ and $\delta^{18}O$. Gaining a comprehensive comprehension of the fundamental process is advantageous to effectively utilise it to discern the varying significance of distinct oxidation pathways, namely gas phase reactions and aerosol reactions, amidst fluctuating environmental variables such as pollution, volcanic occurrences and climate change (Baskaran, 2011a, b).

6.5 *Chlorine*

Chlorine comprises two stable isotopes, ^{35}Cl and ^{37}Cl, constituting relative abundances of 75.78% and 24.22%, respectively. It can react with other metals to form chloride salts. Notably, all these chloride salts—aside from silver chloride (AgCl)—are soluble in water. Solubility in water is a characteristic of inorganic chloride salts that permits their penetration into the hydrosphere. Chloride ions are distributed throughout the hydrosphere's numerous sub-compartments, linked by the hydrological cycle, by several processes. One of the primary sources of atmospheric chlorine is the aerosol produced by sea spray. Much chlorine is trapped in the aerosol and then released back into the ocean through dry and wet deposition. Hydrogen chloride and other inorganic chlorinated compounds are formed when part of chlorine passes through changes for which the mechanisms are not fully understood. Chlorine atoms are produced when these substances interact with the hydroxyl radicals in the air. Subsequently, they react with volatile organic compounds, forming organic chlorines (Cincinelli et al., 2012).

A scientific investigation conducted by Volpe and Spivack (1994) examined the composition of bulk marine aerosols in the atmosphere, revealing a range of values spanning 2‰. Additionally, two separate studies by Liu et al. (2008) and Sun et al. (2004) focused on measuring dissolved Cl^- in rainwater, yielding values ranging from -4‰ to 1.8‰. These findings suggest that the isotopic composition of atmospheric Cl may exhibit heterogeneity. The primary contributor of chlorine (Cl) in the atmosphere is marine aerosols, which are introduced into the air through ocean bubble-breaking or wave-crest tearing mechanisms (Erickson and Duce, 1988). Most atmospheric chlorine (Cl) is promptly deposited back into the Earth's surface through wet or dry deposition mechanisms. The study conducted by Volpe and Spivack (1994) aimed to investigate the stable isotopic composition of marine sea-salt aerosol particles to estimate the atmospheric fluxes of chlorine.

The $\delta^{37}Cl$ values of marine aerosols collected from Bermuda exhibited a positive correlation with the Cl deficit, indicating a fractionation of Cl isotopes during the volatilisation of Cl from aerosols. In volatilisation studies, the observed isotopic fractionation between HCl and NaCl-saturated brines was approximately -3‰, consistent with the isotopic fractionation seen in marine aerosols. This similarity suggests that the most probable mechanism is the acidification of sea-salt aerosols and the subsequent formation of HCl depleted in 37Cl. In a comparable study, Eggenkamp

(1994) documented that the interaction between NaCl brine and H_2SO_4 leads to the enrichment of ^{37}Cl in the evaporate, suggesting the liberation of HCl gas depleted in ^{37}Cl. The ^{37}Cl fractionation factors determined by Schauble et al. (2003) and the experimental data obtained by Howald (1960) suggest that HCl is expected to exhibit a slight enrichment of ^{37}Cl, relative to NaCl, at equilibrium. The isotopic composition of chloride ions (Cl^-) in dissolved form in precipitation indicates a depletion of $\delta^{37}Cl$ compared to seawater, with a maximum deviation of 3.5‰. This observation implies that the atmospheric reservoir of hydrogen chloride (HCl) exhibits a deficiency in the isotopic composition of chlorine-37 ($\delta^{37}Cl$), indicating that its formation is predominantly attributed to the acidification of marine aerosols in the atmosphere. Furthermore, the process of Cl isotope fractionation is mainly governed by kinetic factors and may be influenced by the concentrations of HNO_3 and H_2SO_4 present in the atmosphere.

6.6 *Sulphur*

The isotopic composition of stable sulphur entering the atmosphere exhibits considerable variability. Oxidation of SO_2 to sulphate shows a notable inverse impact, enriching the heavy ^{34}S isotope in the sulphate product. This inverse effect is attributed to an equilibrium step involving the conversion of SO_2 to HSO_3^- before the final oxidation to sulphate takes place.

Sulphur dioxide is a significant industrial contaminant within the Earth's atmosphere. Nevertheless, the origin of sulphur is often challenging to determine, especially when considering regional contexts with multiple coexisting sulphur sources. Using geochemical tracers, specifically the stable sulphur ($\delta^{34}S$) and oxygen ($\delta^{18}O$) isotopic compositions of sulphates in precipitation has proven effective in finding different atmospheric sulphur sources.

Sulphur in the atmosphere exists in diverse states, encompassing gaseous, liquid, and solid forms. These compounds can originate from natural and human-induced sources, such as sea spray, sulphur-reducing bacteria, volcanic emissions, the burning of fossil fuels, and various industrial activities (Newman et al., 1991; Lim et al., 2014). The analysis of pollutant concentrations and isotopic composition, with a particular focus on SO_2, SO_4, and S-aerosols, yields valuable insights into identifying sources of sulphur in the atmosphere (Zhang et al., 2010).

The sulphur isotope in sulphate in precipitation generally indicates the source composition (Harris et al., 2012). On the other hand, the oxygen isotope in sulphate shows the oxygen isotope composition of atmospheric moisture and the prevailing temperature during sulphate synthesis (Holt and Kumar, 1991). After its synthesis, the stable isotopic compositions of sulphur and oxygen in sulphate remain unchanged and do not experience any additional fractionation or isotope exchange. As a result, the original signature from the time of sulphate formation is preserved. Therefore, the utilisation of stable sulphur and oxygen isotope analyses in environmental research might facilitate the identification of the source (natural versus anthropogenic) and transformation mechanisms (inputs versus removal) of S-compounds in the atmosphere (Tichomirowa et al., 2007). Some previous research has effectively utilised the stable isotopic compositions of sulphur and oxygen to monitor the sources and origins of (i) sulphate in precipitation (Tichomirowa et al., 2007; Lim et al., 2014); (ii) sulphur dioxide in the atmosphere (Novák et al., 2000; Zhang et al., 2010); and (iii) aerosol particles (Norman et al., 2006; Sinha et al., 2008).

Yan et al. (2020) indicate that coal burning remains a significant source of sulphate in rainwater despite its substantial reduction of SO_2 emissions during the last twenty years. A negative correlation exists between the $\delta^{34}S$-sulphate and $\delta^{18}O$-sulphate levels in rainfall, with notable seasonal variations. The observed variations in the isotopic composition of sulphate ($\delta^{34}S$ and $\delta^{18}O$) have been attributed to seasonal fluctuations in dimethyl sulphide and coal combustion sources. Additionally, the seasonal trends in $\delta^{18}O$-sulphate align with those of $\delta^{18}O$-water, suggesting that sulphate formation in rainwater primarily occurs through the heterogeneous oxidation of SO_2. Yan et al. (2020) observed a notable rise in the volume-weighted mean $\delta^{34}S$ values of sulphate in rainwater within Guizhou province from 2001 to 2018 from the prior studies of Xiao and Liu (2002), Liu (2007), Xiao et al. (2009, 2014). This trend suggests a decrease in the emission of ^{34}S-depleted SO_2 resulting from coal combustion during this specific time frame. Moreover, the concurrent variations in $\delta^{34}S$ values, sulphate concentration, and pH values of rainwater indicate a decline in the occurrence of acid rain in Guizhou province during the previous twenty-year period. This trend can be attributed to the adoption of emission reduction strategies in Guizhou province.

6.7 Calcium

Identifying the primary source of calcium in precipitation holds significant importance in studying the acid rain phenomenon. The three primary sources of calcium can be traced back to rainwater. The initial source of calcium (Ca) is anthropogenic, resulting from human activities such as traffic emissions, cement production, and similar processes. The second source pertains to the process of carbonate weathering. The third aspect pertains to the long-range transportation of soil particles (Han et al., 2010). Sanusi et al. (1996) found that the construction and building activities might cause significant turbulences, leading to the dispersion of aerosols in the atmosphere.

Calcium (Ca) is a prominent cation in rainwater and primary aerosol reservoirs. Therefore, the isotopic composition of calcium can be utilised to provide insights into the sources of aerosols. The relationship between $\delta^{44/40}Ca$ and 1/Ca (the reciprocal of calcium concentration in μmol/L) is represented as a linear trend in a plot, as observed in the study conducted by Langmuir et al. in 1978. There are three distinct components of calcium (Ca) present in rainwater. The first component, denoted as C, exhibits a high abundance of calcium but a low $\delta^{44/40}Ca$ value of approximately 0.5‰. The second component, marked as S, is characterised by a significantly low calcium concentration and an intermediate $\delta^{44/40}Ca$ value of around 0.8‰. Lastly, the third component, denoted as A, displays a medium calcium concentration but the highest $\delta^{44}/^{40}Ca$ value of approximately 1.2‰. By comparing the concentrations of these components and the concentrations of calcium and the isotope compositions of established reservoirs, it may be inferred that component S is most likely silicate dust, whereas component C is most likely carbonate dust.

6.8 Strontium

The variation in strontium isotope ratios among different sources can be attributed to the varying concentrations of ^{87}Rb, the radiogenic precursor of ^{87}Sr, among these sources. Using the Sr isotope system in conjunction with Sr concentrations enables the examination of the amalgamation of distinct Sr sources (Negrel et al., 2007).

The stable Sr isotope has demonstrated its potential for identifying the source of base cations. This is due to its stability throughout surficial processes and its variation across different rock types, with distinct Rb/Sr values and geological periods (Capo et al., 1998). Nevertheless, the current knowledge regarding the $^{87}Sr/^{86}Sr$ ratio in rainwater is limited. Hence, researchers are posing a potential constraint on the application of Sr isotopes in the field of atmospheric geochemistry. Strontium (Sr) and Calcium (Ca) demonstrate comparable geochemical characteristics in natural environments.

Consequently, it is commonly anticipated that Sr isotopes will determine the origins of base cations, specifically Ca, in rainwater (Nakano and Tanaka, 1997). Multiple authors have reported on the strontium isotope ratios found in rainfall from various regions across the globe (Graustein and Armstrong, 1983; Andersson et al., 1994; Dupre et al., 1994; Negrel and Roy, 1998). The isotopic signature of strontium from different sources exhibits significant variations, enabling the differentiation of diverse aerosol sources (Herut et al., 1993; Aberg, 1995).

Numerous studies have extensively documented the Sr isotopic compositions $^{87}Sr/^{86}Sr$, observed in rainwater worldwide, ranging from 0.7079 to 0.7200 (Andersson et al., 1994). The correlation between $^{87}Sr/^{86}Sr$ or $^{87}Sr/^{86}Sr$ and 1/Sr (the reciprocal of Sr concentration in μmol/litre) necessitates combining three distinct components. Based on the existing Sr reservoirs, the component exhibiting elevated $^{87}Sr/^{86}Sr$ ratios (~0.717) and elevated $\delta^{88}/^{86}Sr$ values (~0.32‰), along with moderate Sr content, is more likely attributed to silicate soil dust. Conversely, the component displaying reduced $^{87}Sr/^{86}Sr$ ratios (~0.708) and lessened $^{88}Sr/^{86}Sr$ values (< ~0.26‰), accompanied by a high Sr concentration, is more likely associated with carbonate soil dust. The Sr isotope properties of these two endmembers exhibit similarities to those previously documented by Pearce et al. (2015a, b).

The rainwater collected in Guiyang City exhibits variations in both stable ($\delta^{44}/^{40}Ca$ and $^{88}Sr/^{86}Sr$) and radiogenic ($^{87}Sr/^{86}Sr$) isotope compositions. These fluctuations are correlated with the inverse of Sr or Ca content and the ratio of $(NH_4^{++}K^+)/$

($Ca_2^{++}Mg_2^{+}$). The observed changes in question indicate the impacts of human activities on the process of dissolving silicate and carbonate aerosols in rainwater, as discussed by Han et al. in their 2019 study.

Soil dust is commonly regarded as a natural reservoir of rainfall strontium (Sr) due to its widespread presence on the Earth's surface and its propensity to be entrained into the atmosphere by wind. In addition, this component has a prominent characteristic of having a greater $^{87}Sr/^{86}Sr$ ratio, typically exceeding 0.710, and a comparatively lower quantity of strontium. This feature plays a crucial role in regulating the strontium levels present in rainfall. Several seminal investigations have substantiated this assertion (Han et al., 2010; Xu et al., 2015). The variability of the $^{87}Sr/^{86}Sr$ ratios is seen with varying anthropogenic sources such as high energy usage, agricultural output, and industrial expansion. An illustrative case is presented in the research conducted by Widory et al. (2010), which demonstrates that air particles resulting from coal combustion exhibit an $^{87}Sr/^{86}Sr$ ratio of 0.70924. Similarly, cement factories display a value of 0.71029 for this ratio, while smelters show the highest radiogenic value of 0.71206. According to Rao et al. (2017), the ratios of $^{87}Sr/^{86}Sr$ exhibit a range of values for different sources. For fertiliser, the range is between 0.7079 and 0.7087. Automobile exhaust shows ratios ranging from 0.7077 to 0.7083. Urban heating sources exhibit ratios between 0.7083 to 0.7335. Incinerators, on the other hand, display ratios ranging from 0.7097 to 0.7100.

6.9 *Lead*

Lead (Pb) is a significant subject of discussion in environmental studies due to its anticipated detrimental effects and distinctive association with specific sources of atmospheric pollutants. In recent years, burning coal has become a significant source of lead (Pb) in aerosols in urban and rural areas. The isotope ratios of $^{206}Pb/^{207}Pb$ in gasoline are less than those of diesel soot (1.159), with values of 1.089. The recommended method for estimating the upper limit of aerosol Pb isotopic compositions considers Pb compounds from three different sources: emissions from a nearby chemical waste incinerator, a metal processing facility, and a prominent natural background. A solid waste incinerator's 206Pb/207Pb ratios are 1.152, but those of a steel factory range from 1.151 to 1.152. The Pb isotope ratio found in diesel fuel is similar to these values. Pb-isotope ratios help locate natural dust sources and determine their dispersion patterns. Furthermore, Geagea et al. (2008) have demonstrated that Pb isotope ratios can be a highly effective approach for monitoring and identifying anthropogenic atmospheric pollution. Despite implementing policies aimed at discontinuing the use of leaded fuel in many nations, there is compelling evidence suggesting that Pb continues to be a prevalent contaminant in the atmosphere (Díaz-Somoano et al., 2009).

6.10 Radionuclides

Three different types of radionuclides have been widely used as atmospheric tracers. These categories comprise radionuclides created by humans through nuclear testing, naturally occurring radionuclides like ^{222}Rn, and cosmogenic radionuclides produced by cosmic rays. Baskaran (2011a, b) studied ^{222}Rn and its offspring, ^{210}Pb, ^{210}Bi, and ^{210}Po, to shed light on the origins, residence times, removal rates, and velocities of aerosol deposition. The atmospheric anthropogenic radionuclides are products of nuclear weapon tests and accidents. These radionuclides have irregular deposition patterns and temporal fluctuations, making them temporary tracers. Conversely, cosmogenic nuclides reliably introduce nuclides that serve as tracers for examining the features of the macroscopic movement of the atmosphere. Radiation from cosmic rays produces a diversity of radionuclides, each with unique chemical properties and half-lives ranging from a few hours to millions of years. Therefore, it is possible to investigate quantitatively air transport/mixing processes and aerosol scavenging through precipitation from hours to years (Lal and Peters, 1967).

The radionuclides in the atmosphere exhibit a quick tendency to adhere to aerosols of minimal size. The variability in ground-level air is influenced by aerosol behaviour, which encompasses a spectrum of potential alterations. Comprehending these processes enables scientists to assess and forecast the global dispersion of contaminants. The approach is precious in evaluating and addressing the possible consequences of unintended emissions of radioactive isotopes from nuclear facilities.

The environment labs of the International Atomic Energy Agency (IAEA) employ large-volume pumps and collectors for concentrating aerosols and capturing dry deposition. This enables the laboratories to assess the number of detectable radionuclides in the atmosphere and monitor any fluctuations in their deposition. Marine sediment traps are employed to investigate the influence of air fallout on the composition of sinking particulate matter within marine ecosystems.

To enhance comprehension of phenomena occurring in the lower troposphere and the mechanisms by which pollutants are transported from the atmosphere to terrestrial and marine ecosystems, researchers from IAEA investigate the rates at which radionuclides, including beryllium-7, caesium-137, and lead-210, are exchanged. These phenomena are generated through the impact of cosmic rays (known as cosmogenic) or human activity (referred to as anthropogenic). A considerable fraction of ^{10}Be is due to fine dust in precipitation, but the contribution of ^{7}Be from terrestrially originated atmospheric dust is negligible. For this reason, the ^{7}Be/^{10}Be ratio has been used as a tracer.

Tanaka and Turekian (1991) estimated ^{35}S in precipitation, aerosol SO_4, and gaseous SO_2 to determine the time constant related to aerosol deposition, in-cloud scavenging, and oxidation. The short half-life of Index elements like titanium (Ti) and aluminium (Al) makes them suitable for quantifying dust particles. Tracing atmospheric halides and chlorinated hydrocarbons is successful when using combination isotopes of iodine-129 and chlorine-36. Seawater recycling in the atmosphere may now be measured thanks to the isotopes ^{36}Cl and ^{22}Na, which have shown to be trustworthy tracers of atmospheric chloride.

7 Conclusions

Isotope tracers offer a novel perspective on atmospheric contaminants' formation, transit, and dispersion. By providing a greater understanding of the underlying mechanisms, the isotopic composition of aerosols and rainfall can augment the creation of prediction models in atmospheric chemistry. Different geographical and temporal dimensions indicate variations in the mechanisms that control the isotopic disproportions between aerosols and precipitation. The isotopic compositions of precipitation that accumulate over extended periods and multiple episodes are frequently associated with temperature, evaporation, and condensation. Discrete events like flow pathways, mixing rate, and air stability significantly impact the isotopic composition of a larger-scale hydrologic system.

The usefulness of natural abundance isotope signatures in tracking ecosystem processes and clarifying patterns and mechanisms at the individual organism level is illustrated by this review. More research is necessary to fully explore the possibility of using stable isotopes to address various environmental issues strategically. A workable method for differentiating atmospheric phenomena and pinpointing contamination sources combines isotope ratio studies with chromatography, computer modelling, and aerosol mass spectrometry. This will improve our understanding of several atmospheric chemical processes, where precipitation and aerosol have a significant impact that may be used to plan future atmospheric isotope chemistry-related research projects.

References

Aberg G (1995) The use of natural strontium isotopes as tracers in environmental studies. Water Air Soil Pollut 79:309–322

Aguilera J, Whigham LD (2018) Using the $^{13}C/^{12}C$ carbon isotope ratio to characterise the emission sources of airborne particulate matter: a review of literature. Isot Environ Health Stud 54:573–587

Andersson PS, Wasserburg GJ, Ingri J, Stordal MC (1994) Strontium, dissolved and particulate loads in fresh and brackish waters: the Baltic Sea and Mississippi Delta. Earth Planet Sci Lett 124(1–4):195–210

Andreae MO, Merlet P (2001) Emission of trace gases and aerosols from biomass burning. Global Biogeochem Cycles 15(4):955–966

Araguás-Araguás L, Froehlich K, Rozanski K (2000) Deuterium and oxygen-18 isotope composition of precipitation and atmospheric moisture. Hydrol Process 14(8):1341–1355

Baskaran M (2011) Po-210 and Pb-210 as atmospheric tracers and global atmospheric Pb-210 fallout: a review. J Environ Radioact 102:500–513

Baskaran M (ed) (2011) Handbook of environmental isotope geochemistry. Springer Science & Business Media
Beck N, Münnich KO (1988) Freezing of water: Isotopic fractionation. Chem Geol 70:168
Bilke S, Mosandl A (2002) Measurements by gas chromatography/pyrolysis/mass spectrometry: fundamental conditions in 2H/1H isotope ratio analysis. Rapid Commun Mass Spectrom 16:468–472
Calvert JG, Lazrus A, Kok GL, Heikes BG, Walega JG, Lind J, Cantrell CA (1985) Chemical mechanisms of acid generation in the troposphere. Nature 317(6032):27–35
Calvo A, Alves C, Castro A et al (2013) Research on aerosol sources and chemical composition: past, current and emerging issues. Atmos Res 120:1–28
Capo RC, Stewart BW, Chadwick OA (1998) Strontium isotopes as tracers of ecosystem processes: theory and methods. Geoderma 82(1–3):197–225
Ceburnis D, Garbaras A, Szidat S, Rinaldi M, Fahrni S, Perron N, Wacker L, Leinert S, Remeikis V, Facchini MC, Prévôt AS, Jennings SG, Ramonet M, O'Dowd CD (2011) Quantification of the carbonaceous matter origin in submicron marine aerosol by 13 C and 14 C isotope analysis. Atmos Chem Phys 11(16):8593–8606
Cincinelli A, Pieri F, Zhang Y, Seed M, Jones KC (2012) Compound Specific Isotope Analysis (CSIA) for chlorine and bromine: a review of techniques and applications to elucidate environmental sources and processes. Environ Pollut 169:112–127. https://doi.org/10.1016/j.envpol.2012.05.006
Court J, Goldsack R, Ferrari L et al (1981) The use of carbon isotopes in identifying urban air particulate sources. Clean Air 15(1):6–11
Craig H (1961) Isotopic variations in meteoric waters. Science 133(3465):1702–1703
Dansgaard W (1964) Stable isotopes in precipitation. Tellus 16(4):436–468
Delmas R, Serça D, Jambert C (1997) Global inventory of NOx sources. Nutr Cycl Agroecosyst 48:51–60
Díaz-Somoano M, Kylander ME, López-Antón MA, Suárez-Ruiz I, Martínez Tarazona MR, Ferrat M, Kober B, Weiss DJ (2009) Stable lead isotope compositions in selected coals from around the world and implications for present day aerosol source tracing. Environ Sci Technol 43(4):1078–1085. https://doi.org/10.1021/es801818r
Dupré B, Negrel P, Seimbille F, Allegre CJ (1994) 87Sr/86Sr ratio variation during a rain event. Atmos Environ 28(4):617–620
Durka W, Schulze ED, Gebauer G, Voerkeliust S (1994) Effects of forest decline on uptake and leaching of deposited nitrate determined from 15N and 18O measurements. Nature 372(6508):765–767
Dusek U, Meusinger C, Oyama B, Ramon W, de Wilde PA, Holzinger R, Röckmann T (2013) A thermal desorption system for measuring $\delta^{13}C$ ratios on organic aerosol. J Aerosol Sci 66:72–82. https://doi.org/10.1016/j.jaerosci.2013.08.005
Eggenkamp HGM (1994) The geochemistry of chlorine isotopes. Ph.D. thesis, Utrecht University
Ehleringer JR, Rundel PW (1989) Stable isotopes: history, units, and instrumentation. In: Rundel PW, Ehleringer JR, Nagy KA (eds) Stable isotopes in ecological research. Ecological studies, vol 68. Springer, New York, pp 1–15
Elliott EM, Kendall C, Boyer EW, Burns DA, Lear GG, Golden HE, Harlin K, Bytnerowicz A, Butler TJ, Glatz R (2009) Dual nitrate isotopes in dry deposition: utility for partitioning NOx source contributions to landscape nitrogen deposition. J Geophys Res: Biogeosci 114(G4):1–15
Erickson DJ, Duce RA (1988) On the global flux of atmospheric sea salt. J Geophys Res – Oceans 93:14079–14088
Felix JD, Elliott EM, Shaw SL (2012) Nitrogen isotopic composition of coal-fired power plant NOx: influence of emission controls and implications for global emission inventories. Environ Sci Technol 46(6):3528–3535
Fenech C, Rock L, Nolan K, Tobin J, Morrissey A (2012) The potential for a suite of isotope and chemical markers to differentiate sources of nitrate contamination: a review. Water Res 46(7):2023–2041. https://doi.org/10.1016/j.watres.2012.01.044

Fibiger DL, Hastings MG (2016) First measurements of the nitrogen isotopic composition of NOx from biomass burning. Environ Sci Technol 50(21):11569–11574

Galloway JN, Dentener FJ, Capone DG, Boyer EW, Howarth RW, Seitzinger SP, Asner GP, Cleveland CC, Green PA, Holland EA, Karl DM, MichaelsAF PJH, Vöosmarty CJ (2004) Nitrogen cycles: past, present, and future. Biogeochemistry 70:153–226

Galloway JN, Townsend AR, Erisman JW, Bekunda M, Cai Z, Freney JR, Martinelli LA, Seitzinger SP, Sutton MA (2008) Transformation of the nitrogen cycle: recent trends, questions, and potential solutions. Science 320(5878):889–892

Garbarienė I, Šapolaitė J, Garbaras A et al (2016) Origin identification of carbonaceous aerosol particles by carbon isotope ratio analysis. Aerosol Air Qual Res 16(6):1356–1365

Geagea ML, Stille P, Gauthier-Lafaye F, Millet M (2008) Tracing of industrial aerosol sources in an urban environment using Pb, Sr, and Nd isotopes. Environ Sci Technol 42(3):692–698. https://doi.org/10.1021/es071704c

Goldstein AH, Shaw SL (2003) Isotopes of volatile organic compounds: an emerging approach for studying atmospheric budgets and chemistry. Chem Rev 103:5025–5048

Graustein WC, Armstrong RL (1983) The use of strontium-87/strontium-86 ratios to measure atmospheric transport into forested watersheds. Science 219(4582):289–292

Gröning M (2004) International stable isotope reference materials. In: de Groot P (ed) Handbook of stable isotope analytical techniques, vol I. Elsevier, Amsterdam, pp 874–906

Guha T, Lin CT, Bhattacharya SK, Mahajan AS, Ou-Yang CF, Lan YP, Hsu SC, Liang MC (2017) Isotopic ratios of nitrate in aerosol samples from Mt. Lulin, a high-altitude station in Central Taiwan. Atmos Environ 154:53–69

Han G, Song Z, Tang Y, Wu Q, Wang Z (2019) Ca and Sr isotope compositions of rainwater from Guiyang city, Southwest China: implication for the sources of atmospheric aerosols and their seasonal variations. Atmos Environ 214:116854

Han G, Tang Y, Wu Q, Tan Q (2010) Chemical and strontium isotope characterization of rainwater in karst virgin forest, Southwest China. Atmos Environ 44(2):174–181. https://doi.org/10.1016/j.atmosenv.2009.10.019

Harris E, Sinha B, Hoppe P, Crowley JN, Ono S, Foley S (2012) Sulfur isotope fractionation during oxidation of sulfur dioxide: Gas-phase oxidation by OH radicals and aqueous oxidation by H_2O_2, O_3 and iron catalysis. Atmos Chem Phys 12(1):407–423

Heaton THE (1987) 15N14N ratios of nitrate and ammonium in rain at Pretoria, South Africa. Atmos Environ (1967) 21(4):843–852

Heaton THE (1990) 15N/14N ratios of NO_x from vehicle engines and coal-fired power stations. Tellus B 42(3):304–307

Heintzenberg J, Hermann M, Theiss D (2003) Out of Africa: high aerosol concentrations in the upper troposphere over Africa. Atmos Chem Phys 3(4):1191–1198

Herut B, Starinsky A, Katz A (1993) Strontium in rainwater from Israel: sources, isotopes and chemistry. Earth Planet Sci Lett 120(1–2):77–84

Hildemann LM, Klinedinst DB, Klouda GA et al (1994) Sources of urban contemporary carbon aerosol. Environ Sci Technol 28(9):1565–1576

Ho K, Lee S, Cao J et al (2006) Variability of organic and elemental carbon, water soluble organic carbon, and isotopes in Hong Kong. Atmos Chem Phys 6(12):4569–4576

Holt BD, Kumar R (1991) Oxygen isotope fractionation for understanding the sulphur cycle. Stable Isot: Nat Anthropog Sulphur Environ: SCOPE 43:27–41

Howald RA (1960) Ion pairs. I. Isotope effects shown by chloride solutions in glacial acetic acid 1. J Am Chem Soc 82:20–24

Hsu SC, Huh CA, Lin CY, Chen WN, Mahowald NM, Liu SC, Chou CC, Liang MC, Tsai CJ, Lin FJ, Chen JP, Huang YT (2012) Dust transport from non-East Asian sources to the North Pacific. Geophys Res Lett 39(12):12804

Intergovernmental Panel on Climate Change (IPCC), Climate Change (2007) The physical science basis. Cambridge University Press, UK, p 2007

Jacob DJ (1999) Introduction to atmospheric chemistry. Princeton University Press

Jia W, Peng P, Liu J (2008) Gas chromatography flow rates for determining deuterium/hydrogen ratios of natural gas by gas chromatography/high-temperature conversion/isotope ratio mass spectrometry. Rapid Commun Mass Spectrom 22:2521–2525

Jin Z, Qian L, Shi Y, Fu G, Li G, Li F (2021) Quantifying major NO_x sources of aerosol nitrate in Hangzhou, China, by using stable isotopes and a Bayesian isotope mixing model. Atmos Environ 244:117979

Kampa M, Castanas E (2008) Human health effects of air pollution. Environ Pollut 151(2):362–367

Kendall C (1998) Tracing nitrogen sources and cycling in catchments. In: Kendall C, McDonnell JJ (eds) Isotope tracers in catchment hydrology. Elsevier Science B.V, Amsterdam, pp 519–576

Kim K-H, Kabir E, Kabir S (2015) A review on the human health impact of airborne particulate matter. Environ Int 74:136–143

Lal D, Peters B (1967) Cosmic ray produced radioactivity on the Earth. Kosmische Strahlung II/ Cosmic Rays II. Springer, Berlin, pp 551–612

Langmuir CH, Vocke RD Jr, Hanson GN, Hart SR (1978) A general mixing equation with applications to Icelandic basalts. Earth Planet Sci Lett 37(3):380–392

Li D, Wang X (2008) Nitrogen isotopic signature of soil-released nitric oxide (NO) after fertilizer application. Atmos Environ 42(19):4747–4754

Lim C, Jang J, Lee I, Kim G, Lee SM, Kim Y, Kim H, Kaufman AJ (2014) Sulfur isotope and chemical compositions of the wet precipitation in two major urban areas, Seoul and Busan, Korea. J Asian Earth Sci 79:415–425

Liu CQ (2007) Bio-geochemical processes and cycling of nutrients in the earth's surface: erosion of karstic catchment and nutrients cycling in Southwest China. Science Press, Beijing, p 1e608 (in Chinese)

Liu CQ, Lang YC, Satake H, Wu J, Li SL (2008) Identification of anthropogenic and natural inputs of sulfate and chloride into the karstic ground water of Guiyang, SW China: combined $\delta^{37}Cl$ and $\delta^{34}S$ approach. Environ Sci Technol 42:5421–5427

Liu W, Yang H, Li L (2006) Hydrogen isotopic compositions of n-alkanes from terrestrial plants correlate with their ecological life forms. Oecologia 150:330–338

Lu H, Li C, Sun Y, Peng P, Xiao Z (2005) Hydrogen isotopic compositions, distributions and source signals of individual n-alkanes for some typical crude oils in Lunnan Oilfield, Tarim Basin, NW China. Sci China Ser D Earth Sci 48:1220–1226

Michalski G, Scott Z, Kabiling M, Thiemens MH (2003) First measurements and modeling of $\Delta^{17}O$ in atmospheric nitrate. Geophys Res Lett 30(16)

Mkoma SL, Kawamura K, Tachibana E, Fu P (2014) Stable carbon and nitrogen isotopic compositions of tropical atmospheric aerosols: Sources and contribution from burning of C3 and C4 plants to organic aerosols. Tellus B: Chem Phys Meteorol 66(1):20176

Nakano T, Tanaka TI (1997) Strontium isotope constraints on the seasonal variation of the provenance of base cations in rain water at Kawakami, central Japan. Atmos Environ 31(24):4237–4245

National Research Council (NRC) (2004) Research priorities for airborne particulate matter, IV continuing research progress. National Academy Press, Washington, UDA

Negrel P, Roy S (1998) Chemistry of rainwater in the Massif Central (France): a strontium isotope and major element study. Appl Geochem 13(8):941–952

Negrel P, Guerrot C, Millot R (2007) Chemical and strontium isotope characterization of rainwater in France: influence of sources and hydrogeochemical implications. Isot Environ Health Stud 43(3):179–196

Nestler A, Berglund M, Accoe F, Duta S, Xue D, Boeckx P, Taylor P (2011) Isotopes for improved management of nitrate pollution in aqueous resources: review of surface water field studies. Environ Sci Pollut Res 18(4):519–533

Newman L, Krouse HR, Grinenko VA (1991) Sulphur isotope variations in the atmosphere. In: Stable isotopes: natural and anthropogenic Sulphur in the environment, UN

Nlend B, Celle-Jeanton H, Risi C, Pohl B, Huneau F, Boum-Nkot SN, Seze G, Roucou P, Camberlin P, Etame J, Ketchemen-Tandia B (2020) Identification of processes that control the stable isotope composition of rainwater in the humid tropical West-Central Africa. J Hydrol 584:124650

Norman AL, Anlauf K, Hayden K, Thompson B, Brook JR, Li SM, Bottenheim J (2006) Aerosol sulphate and its oxidation on the Pacific NW coast: S and O isotopes in $PM_{2.5}$. Atmos Environ 40(15):2676–2689

Novák M, Kirchner JW, Groscheová H, Havel M, Černý J, Krejčí R, Buzek F (2000) Sulfur isotope dynamics in two Central European watersheds affected by high atmospheric deposition of SO_x. Geochim Cosmochim Acta 64(3):367–383

Pavuluri CM, Kawamura K, Tachibana E, Swaminathan T (2010) Elevated nitrogen isotope ratios of tropical Indian aerosols from Chennai: implication for the origins of aerosol nitrogen in South and Southeast Asia. Atmos Environ 44(29):3597–3604

Pearce CP, Parkinson IJ, Gaillardet J, Chetelat B, Burton KW (2015a) Characterising the stable ($\delta^{88/86}Sr$) and radiogenic ($^{87}Sr/^{86}Sr$) isotopic composition of strontium in rainwater. Chem Geol 409:54–60

Pearce CR, Parkinson IJ, Gaillardet J, Charlier BLA, Mokadem F, Burton KW (2015b) Reassessing the stable ($\delta^{88/86}Sr$) and radiogenic ($^{87}Sr/^{86}Sr$) strontium isotopic composition of marine inputs. Geochim Cosmochem Acta 157:125–146

Poulson SR, Drever JI (1999) Stable isotope (C, Cl, and H) fractionation during vaporization of trichloroethylene. Environ Sci Technol 33:3689–3694

Prather MJ, Ehhalt DH, Dentener FJ, Derwent RG, Dlugokencky EJ, Holland E, Isaksen IS, Katima JH, Kirchhoff, Matson PA, Midgley PM, Wang M, Berntsen TK, Bey I, Brasseur GP, Buja L, Pitari GM (2001) Contribution to Chapter 4: Atmospheric chemistry and greenhouse gases. The IPCC Third Assessment Report Climate Change

Rao W, Han G, Tan H, Jiang S (2015) Chemical and Sr isotopic compositions of rainwater on the Ordos Desert Plateau, Northwest China. Environ Earth Sci 74:5759–5771

Rao W, Han G, Tan H, Jin K, Wang S, Chen T (2017) Chemical and Sr isotopic characteristics of rainwater on the Alxa Desert Plateau, North China: implication for air quality and ion sources. Atmos Res 193:163–172

Rice AL, Gotoh AA, Ajie HO, Tyler SC (2001) High-precision continuous-flow measurement of $\delta^{13}C$ and δD of atmospheric CH_4. Anal Chem 73:4104–4110

Rice AL, Quay PD (2006) Isotopic analysis of atmospheric formaldehyde by gas chromatography isotope ratio mass spectrometry. Anal Chem 78:6320–6326

Rudolph J, Czuba E, Norman AL, Huang L, Ernst D (2002) Stable carbon isotope composition of nonmethane hydrocarbons in emissions from transportation related sources and atmospheric observations in an urban atmosphere. Atmos Environ 36:1173–1181

Sanusi A, Wortham H, Millet M, Mirabel P (1996) Chemical composition of rainwater in eastern France. Atmos Environ 30:59–71

Savarino J, Kaiser J, Morin S, Sigman DM, Thiemens MH (2007) Nitrogen and oxygen isotopic constraints on the origin of atmospheric nitrate in coastal Antarctica. Atmos Chem Phys 7(8):1925–1945

Schauble EA, Rossman GR, Taylor HP (2003) Theoretical estimates of equilibrium chlorine-isotope fractionations. Geochim Cosmochim Acta 67:3267–3281

Schmidt TC, Zwank L, Elsner M, Berg M, Meckenstock RU, Haderlein SB (2004) Compound-specific stable isotope analysis of organic contaminants in natural environments: a critical review of the state of the art, prospects, and future challenges. Anal Bioanal Chem 378(2):283–300

Schumann U, Huntrieser H (2007) The global lightning-induced nitrogen oxides source. Atmos Chem Phys 7(14):3823–3907

Sharma SK, Mandal TK, Shenoy DM et al (2015) Variation of stable carbon and nitrogen isotopic composition of PM10 at urban sites of Indo Gangetic Plain (IGP) of India. Bull Environ Contam Toxicol 95(5):661–669

Sinha BW, Hoppe P, Huth J, Foley S, Andreae MO (2008) Sulfur isotope analyses of individual aerosol particles in the urban aerosol at a central European site (Mainz, Germany). Atmos Chem Phys 8(23):7217–7238

Sun A, Xiao Y, Wang Q, Zhang C, Wei H, Liao B, Li S (2004) The separation and stable isotope measurement of chlorine in low-concentration liquid sample. Chin J Anal Chem 10:1362–1364

Tanaka N, Turekian KK (1991) Use of cosmogenic ^{35}S to determine the rates of removal of atmospheric SO_2. Nature 352(6332):226–228

Tichomirowa M, Haubrich F, Klemm W, Matschullat J (2007) Regional and temporal (1992–2004) evolution of air-borne sulphur isotope composition in Saxony, southeastern Germany, Central Europe. Isot Environ Health Stud 43(4):295–305

Urey HC (1948) Oxygen isotopes in nature and in the laboratory. Science 108(2810):489–496

USEPA (U.S. Environmental Protection Agency) (2008) Draft nanomaterial research strategy. EPA/600/S-08/002. U.S. Environmental Protection Agency, Office of Research and Development, Washington, DC

Vitzthum Von Eckstaedt C, Grice K, Ioppolo-Armanios M, Chidlow G, Jones M (2011a) δD and $\delta^{13}C$ analyses of atmospheric volatile organic compounds by thermal desorption gas chromatography isotope ratio mass spectrometry. J Chromatogr 1218:6511–6517

Vitzthum Von Eckstaedt C, Grice K, Ioppolo-Armanios M, Jones M (2011b) $\delta^{13}C$ and δD of volatile organic compounds in an alumina industry stack emission. Atmos Environ 45:5477–5483

Volpe C, Spivack AJ (1994) Stable chlorine isotopic composition of marine aerosol particles in the western Atlantic Ocean. Geophys Res Lett 21:1161–1164

Walters WW, Goodwin SR, Michalski G (2015) Nitrogen stable isotope composition (δ15N) of vehicle-emitted NO_x. Environ Sci Technol 49(4):2278–2285

Walters WW, Michalski G (2015) Theoretical calculation of nitrogen isotope equilibrium exchange fractionation factors for various NO_y molecules. Geochim Cosmochim Acta 164:284–297

Wang Y, Huang Y (2001) Hydrogen isotope fractionation of low molecular weight n-alkanes during progressive vaporization. Org Geochem 32:991–998

Wang Y, Huang Y (2003) Hydrogen isotope fractionation of petroleum hydrocarbons during vaporization: Implications for assessing artificial and natural remediation of petroleum contamination. Appl Geochem 18:1641–1651

Widory D, Roy S, Le Moullec Y et al (2004) The origin of atmospheric particles in Paris: a view through carbon and lead isotopes. Atmos Environ 38(7):953–961

Widory D, Liu X, Dong S (2010) Isotopes as tracers of sources of lead and strontium in aerosols (TSP & PM2.5) in Beijing. Atmos Environ 44(30):3679–3687

Williams EJ, Guenther A, Fehsenfeldi FC (1992) An inventory of nitric oxide emissions from soils in the United States. J Geophys Res: Atmos 97(D7):7511–7519

Xiao HW, Xiao HY, Long AM, Wang YL, Liu CQ (2014) Sources and meteorological factors that control seasonal variation of δ34S values in rainwater. Atmos Res 149:154–165

Xiao HY, Liu CQ (2002) Sources of nitrogen and sulfur in wet deposition at Guiyang, southwest China. Atmos Environ 36(33):5121–5130

Xiao HY, Tang CG, Xiao HW, Liu XY, Liu CQ (2009) Identifying the change in atmospheric sulfur sources in China using isotopic ratios in mosses. J Geophys Res: Atmos 114(D16)

Xu Z, Wu Y, Liu WJ, Liang CS, Ji J, Zhao T, Zhang X (2015) Chemical composition of rainwater and the acid neutralizing effect at Beijing and Chizhou city, China. Atmos Res 164:278–285

Yamada K, Ozaki Y, Nakagawa F, Tanaka M, Yoshida N (2003) An improved method for measurement of the hydrogen isotope ratio of atmospheric methane and its application to a Japanese urban atmosphere. Atmos Environ 37:1975–1982

Yamamoto S, Kawamura K (2011) Stable hydrogen isotope ratios of n-alkanes in atmospheric aerosols from Okinawa, Japan. Res Org Geochem 27:81–89. https://doi.org/10.20612/rog.27.0_81

Yan Z, Han X, Lang Y, Guo Q, Li S (2020) The abatement of acid rain in Guizhou province, southwestern China: implication from sulfur and oxygen isotopes. Environ Pollut 267:115444

Yeatman SG, Spokes LJ, Dennis PF, Jickells TD (2001a) Can the study of nitrogen isotopic composition in size-segregated aerosol nitrate and ammonium be used to investigate atmospheric processing mechanisms? Atmos Environ 35(7):1337–1345

Yeatman SG, Spokes LJ, Dennis PF, Jickells TD (2001b) Comparisons of aerosol nitrogen isotopic composition at two polluted coastal sites. Atmos Environ 35(7):1307–1320

Yoshimura K (2015) Stable water isotopes in climatology, meteorology, and hydrology: a review. J Meteorol Soc Jpn Ser II 93(5):513–533

Yu Z, Elliott EM (2017) Novel method for nitrogen isotopic analysis of soil-emitted nitric oxide. Environ Sci Technol 51(11):6268–6278

Zhang M, Wang S, Ma G, Zhou H, Fu J (2010) Sulfur isotopic composition and source identification of atmospheric environment in central Zhejiang, China. Sci China Earth Sci 53:1717–1725

Application of Remote Sensing and GIS for the Study of Health Effects of Air Pollution: Challenges and Future Perspectives

Bijaya Kumar Padhi and Pawan Kumar

1 Introduction

When it comes to understanding and relating to the "what", "when," "how," and "why" of the world, remote sensing (RS), and geographic information systems (GIS) are unique kinds of technologies commonly known as "RS-GIS" techniques. GIS facilitates the process by which we can visualize, analyze and understand the data. In contrast, remote sensing is commonly used to collect physical data to integrate into GIS software. Maps and satellite imagery are essential parts of RS-GIS techniques. A GIS draws on ideas and themes from cartography, the academic study of maps, and mapping (Geographic Information System, n.d.).

The tool is beneficial for a variety of academic disciplines, including climatology, which investigates the causes and effects of global warming, epidemiology, which looks for the cause of an outbreak of an infectious disease, political consulting, which creates election campaign strategies, and many other disciplines. The international landscape of healthcare has seen significant changes in recent decades. GIS and related spatial analytic methods offer tools to document and understand how the geographic structure of health care is changing, to analyze how it relates to access and health outcomes, and to consider how health care delivery might be improved. The scope of GIS contributions has expanded significantly in recent years, although it has been used for many years to evaluate healthcare systems. The range of institutions that use GIS includes hospitals, medical facilities, health insurance firms, public health departments, and research groups (McLafferty, 2003).

B. K. Padhi (✉) · P. Kumar
Department of Community Medicine and School of Public Health, Postgraduate Institute of Medical Education and Research, Chandigarh, India
e-mail: bkpadhi@gmail.com

P. K. Padhy et al. (eds.), *Airborne Particulate Matter*,
https://doi.org/10.1007/978-3-031-84408-9_4

1.1 Geographic Information System (GIS)

A GIS is a computer programme that gathers, saves, validates, and displays data about the places on the surface of the earth. GIS can represent streets, structures and plants on maps. Location information is used by GIS software, such as coordinates (latitude and longitude), addresses, or ZIP codes. It provides an attribute table with spreadsheet-based data. A range of data, including digital, photographic, spreadsheet, and cartographic data, may be used in GIS. Cartographic data is used to make maps. Photo interpretation is the process of analysing and evaluating the elements visible in aerial photography utilising photographic data. The GIS may also receive digital data from remote sensing satellites that depicts the location of towns, farms, and other human settlements as well as the usage of land cover. Individuals and organizations can use GIS to better understand spatial links and trends.

The two most common GIS file types are raster and vector. Grids of pixels or cells make up raster formats. Raster formats help store GIS data, such as elevation or satellite imagery. Vector-format polygons are made up of nodes, points and lines. For storing GIS data with distinct borders, such as school districts or streets, vector formats are suitable (Geographic Information System, n.d.; GIS and Public Health at CDC|Cdc.Gov, n.d.).

1.2 Applications of GIS

GIS has been applied in numerous sectors over time due to its advantages. Its potential uses and applications are limitless as technology advances, society becomes more geospatially capable, and more and more individuals rediscover geography and the value of maps. GIS is used to raise public awareness and spread information about the environment, natural resources, potential hazards, and planned urban pathways.

This vital tool is now being actively used worldwide for various applications that many commercially available GIS software packages are becoming more user-friendly and can be used on personal PCs. This tool is utilized in disease mapping, disease control, and public health (GIS Uses in Public Health, 2019). People's data, such as population, income, or level of education, can also be depicted by GIS. It may provide details of the topography. GIS technology enables people to compare the positions of diverse things to ascertain how they relate to one another. For instance, a single map created with GIS might display areas that pollute and areas that are vulnerable to pollution, such as wetlands and rivers, as well as areas where pollution occurs, e.g. industrial areas. Such map might be used to show locations where water supplies are under danger. Despite the fact that using GIS has numerous advantages, it also has certain drawbacks.

1.3 Remote Sensing (RS)

Another tool that can be integrated into GIS is remote sensing. When we hear the term 'remote sensing', we wonder what it means. The term 'remote' refers to a place that is far away, while the term 'sensing' refers to something that is believed, observed, or learned. 'Remote sensing' refers to gathering data on objects from a distance (What Is Remote Sensing?, n.d.). Depending on the signal source utilized to scan the object, remote sensing may be split into two main categories, viz. active and passive. While active remote sensing devices require their light emission source to work, passive rely on reflected light (Types Of Remote Sensing: Technology Changing The World, n.d.). The energy reflected from the earth is measured using a sensor. In order to view Earth and other planetary bodies, NASA deploys remote sensors on satellites and aeroplanes that can identify and capture energy that is emitted or reflected. Since they rely on externally produced data, cartography and GIS and other mapping disciplines are different from remote sensing. Remote sensors provide a global perspective and a wealth of data on earth systems, enabling data-informed decision making based on the current and probable future status of our planet. It has emerged as a separate topic of study because of its transdisciplinary uses and integration with several other scientific and technological domains.

1.4 Applications of Remote Sensing (RS)

Remote sensing can offer basic biophysical information, such as locations, elevation depth, temperature, and moisture content, in well-regulated settings. Additionally, it may gather information on the earth's water cycle, including information on ocean evaporation, atmospheric water vapour, clouds, precipitation, soil moisture, sea and land ice, and snow cover, among other things. Research on alterations in air quality and their connections to climate change is made possible by this. In order to quantify and map air pollution, remote sensing satellite data might be a great resource. This technology is also useful for mapping land cover, urban planning and earth management. Numerous significant decisions are based on data from various satellites, including LANDSAT, MODIS, RADARSAT, TerraSAR-X, SRTM, EOS, ERS and Sentinel. Both technologies, GIS and RS, have certain benefits and drawbacks. Understanding the benefits and drawbacks of these is important to making more effective use (Applications of Remote Sensing in Air Pollution, n.d.).

2 Air Pollution and Public Health

One of the biggest issues for public health is air pollution, which is widespread throughout the world. For health and prosperity, it poses a serious threat. The main impact of air pollution on plants, animals, and people is caused by poisonous gases, dust and smoke. The combustion of fossil fuels, emissions from vehicles, factories, industrial plants, mining and agricultural activities, all contribute to air pollution (Fuller et al., 2022). Pollutants that are deadly to humans and to flora and animals, include carbon monoxide, nitrogen oxide, ozone, aerosols, sulfur dioxide, volatile organic compounds (VOCs), polycyclic aromatic hydrocarbons (PAHs), and particulate matter (PM), which has a varied number but a very small diameter. These pollutants harm the reproductive, cardiovascular, reproductive, and central nervous systems when inhaled. Long-term exposure to pollutants can have negative effects on the lungs' ability to function, lead to lung disease development such as chronic obstructive pulmonary disease (COPD), asthma, bronchitis, and emphysema, and even increase the risk of developing cancer (Manisalidis et al., 2020). Due to overpopulation, unchecked urbanization, and industrialization growth, the problem is that air pollution is particularly severe in developing countries. The use of remote sensing and GIS techniques can be combined with the use of solar, wind and geothermal energy in several ways to reduce pollution (Manisalidis et al., 2020).

2.1 Role of RS-GIS in Air Quality Monitoring

Since air pollution is a serious public health concern, the study of air pollution monitoring, which also includes modelling spatial health-related outcomes, has undergone a paradigm shift (Table 1). In this situation, RS-GIS technology is a preferred tool for air quality management among public health specialists. An approach is to track changes in air quality with the help of satellite sensors positioned in the atmosphere. They can inspect the air from the ground and can give people pertinent information that they can use to control activities that pollute the air. The impact of a gas or aerosol on the atmosphere can be measured using sensors. The station receives the recorded measures, which are then used to improve air quality. The data captured by the sensors is transformed into digital numbers that are applied during data analysis. With the use of remote sensing technology, people can avoid pollutants that cause air quality to decline by tracking them with their sensors. Remote sensing technology can be used to reduce the air pollution that is typically associated with urban areas. To take action against industries that release untreated chemicals into the air, satellite sensors are used to identify those industries and reduce air pollution in metropolitan areas. To find the problem and address it to reduce further pollution, remote sensing tools can be employed. Figure 1 depicts a bibliometric analysis of keyword co-occurrence in RS-GIS research on air pollution and its health effects.

Table 1: Application of RS-GIS techniques in air pollution and health effects research

Author's name	*Country*	*Name of the study*	*Key findings*
Al-Alola S.S. et al. (2022)	Saudi Arabia	Air quality estimation using remote sensing and GIS-spatial technologies along Al-Shamal train pathway, Al-Qurayyat City in Saudi Arabia (Al-Alola et al., 2022)	This study uses satellite data to identify air pollution characteristics and examine how they relate to weather variables such as temperature and wind speed along the Al-Shamal railway route in Al-Qurayyat City, Saudi Arabia. Instead of using conventional measurements from ground stations, the current study looked at using Sentinel-5 Precursor, Moderate Resolution Imaging Spectroradiometer (MODIS), and Modern-Era Retrospective Analysis for Research and Applications, Version 2 (MERRA-2) satellite data to investigate and detect air pollution parameters in the study area. In comparison to the WHO's recommended limits, the study of satellite photos showed an increase in the concentration of air pollution indices, which can have serious detrimental effects on both the environment and human health. With values ranging from 0.0004 to 0.6604, the Pearson correlation coefficient was used to examine the relationship between different air contaminants and temperature and wind speed. A strong linear correlation between air contaminants that were analyzed can be linked to the same source of emission. CO and NO_2 on roads have a high link. NO_2 and CO do not appear to be strongly correlated with SO_2 or AOD, and their correlation coefficients are weaker, pointing to alternative sources of emissions.
Narashid H.N. et al. (2010)	Malaysia	Air quality monitoring using remote sensing and GIS technology (Rohayu Haron Narashid & Wan Mohd Naim Wan Mohd, 2010)	The study revealed that the authors have developed a new technique that combines satellite remote sensing with GIS. Eight (8) continuous air quality monitoring stations (CAQM) data and images from Landsat 7 Enhanced Thematic Mapper Plus (ETM+) are utilised to analyse the link between the digital number (DN) of the thermal infrared band and the two air pollutant characteristics, i.e. CO & PM-10. The idea of "virtual stations" is presented to densify CAQM stations because there are only a limited number of CAQM stations available. The air quality maps of the research region are produced using the kriging interpolation method, which is a feature of GIS software, based on these stations. With fewer CAQM stations, this method enables a more thorough evaluation of air quality in metropolitan areas. A more accurate representation of the distribution of air pollution is produced by maps of pollution concentration produced from virtual stations using the kriging interpolation method. The industrial zones have the highest concentrations of CO and PM_{10} (Shah Alam). Environmental managers and local authorities can use this technique to continuously monitor air quality in urban areas.

(continued)

Table 1: (continued)

Author's name	*Country*	*Name of the study*	*Key findings*
Somvanshi S.S. et al. (2019)	India	Delhi Air Pollution Modelling Using Remote Sensing Technique (Somvanshi et al., 2019)	In this study, they create a GIS model that will make it simple to get air quality data directly from remotely sensed data. The model allows spatially changing interactions between ground observations and other factors that assess air quality inside the multiple linear regression modelling framework. The research uses GIS as a supporting tool to show the potential of Landsat 8 OLI-TIRS for the monitoring of air quality. The study demonstrated that the model, which was created utilising multiple bands of the Landsat 8 satellite image, can accurately estimate particulate matter ($PM_{2.5}$ and PM_{10}) but not gaseous contaminants (NO_x, CO, etc.). Additionally, the model was verified using ground-truth information from 2017.
Nikolov H., Borisova D. (2014)	Bulgaria	A framework for air quality monitoring based on free public data and opensource tools (Nikolov & Borisova, 2014)	In this study, an effort is made to create a working framework based entirely on publicly available data on air quality and a collection of open-source software tools for processing and modelling while taking into account the country's current situation. The National Institute of Meteorology and Hydrology of Bulgaria (NIMH) and the National System of Environmental Monitoring run by the Bulgarian Executive Environmental Agency should be mentioned as two of the main data sources, particularly for larger urban areas, for various types of gases and dust particles. Just to name a few, both authorities provide data on a monthly (and in some cases daily) basis for the concentrations of CO, CO_2, NO_2, SO_2, and fine suspended particles (PM_{10}, $PM_{2.5}$). These data will supplement data from satellite-based sensors, such as the OMI instrument onboard the EOS-Aura satellite and from the TROPOMI instrument payload for the upcoming ESA Sentinel-5P mission, under the framework that has been proposed.
Bai Y. et al. (2016)	China	A Geographically and Temporally Weighted Regression Model for Ground-Level $PM_{2.5}$ Estimation from Satellite-Derived 500 m Resolution AOD (Bai et al., 2016)	The mass concentration of ambient fine particulate matter ($PM_{2.5}$) with an aerodynamic diameter smaller than 2.5 m is used in numerous studies as a crucial indicator to assess air quality. This study proposes a methodology for producing ground-level $PM_{2.5}$ concentrations using satellite-derived 500 m AOD called geographically and temporally weighted regression (GTWR). To represent the spatiotemporal dynamics in the $PM_{2.5}$-AOD relationship, the AOD product GTWR model combines the SARA (simplified high resolution MODIS aerosol retrieval algorithm) with meteorological variables such as the height of the planetary boundary layer (PBLH), relative humidity (RH), wind speed (WS) and temperature (TEMP) extracted from WRF assimilation.

(continued)

Table 1: (continued)

Author's name	*Country*	*Name of the study*	*Key findings*
Akbari M. et al. (2021)	Iran	A GIS-Based System for Real-Time Air Pollution Monitoring and Alerting Based on OGC Sensors Web Enablement Standards (Akbari et al., 2021)	The study revealed that the use of sensor webs is a key strategy for managing and controlling air quality in megacities. The suggested system integrates sensors and their observations while employing them in the geographic data infrastructure by using the SWE framework of OGC, the industry standard in spatial data. The created system has the ability to gather, transport, share and process sensor observations, determine the state of air quality, and report urgent situations in real time. For this, a four-tier architecture structure has been created that includes a sensor, a web service, a logical layer and a presentation layer. The system uses web sensor data to provide a collection of web services using predefined algorithms and subsystems to provide alert information. The created system detects carbon monoxide, ozone, and sulfur dioxide at 20 locations around Tehran and is evaluated through sensor observation. In this way, the system calculates the real-time air quality index and notifies users who have enrolled for the system by email when critical conditions arise.
Ahmad et al. (2006)	Malaysia	Accessibility of Air Quality Over USM Campus Using Remote Sensing and GIS-Technique (Ahmad Jabatan Sains et al., 2006)	To determine the air quality on the campus of USM using spectroradiometer data, this study looked at the sky radiation. The aim is to assess a spectroradiometer's capability to deliver valuable remotely sensed data for investigations into air pollution. Air transmittance measurements were used to calculate the aerosol optical thickness (AOT) values. Using a portable DustTrak Meter, the concentrations of particulate matter smaller than 10 microns (PM_{10}) were measured concurrently with the atmospheric transmittance measurements. The air transmittance was used to calculate aerosol optical thickness (AOT) using the well-known Beer-Lambert-Bouguer law. According to this research, spectroradiometer observations offer helpful remotely sensed information for air quality studies.
He Junliang et al. (2013)	China	Aerosol Indices Derived from MODIS Data to Indicate Aerosol-Induced Air Pollution (He et al., 2014)	In determining aerosol concentration in the atmosphere, assessing the severity of air pollution, and researching their effects on the climate, the aerosol optical depth (AOD) is a crucial quantity. The most and least sensitive MODIS wavebands were determined based on how closely they were related. The newly suggested difference aerosol index, it has been determined, can be used to investigate the degree of aerosol-induced air pollution from MODIS satellite imagery with a fair amount of success.

(continued)

Table 1: (continued)

Author's name	*Country*	*Name of the study*	*Key findings*
Sikder S. et al. (2018)	Bangladesh	Aerosol Optical Thickness (AOT) Assessment Using GIS and Remote Sensing (Sikder & Rubaiyat Islam, 2018)	One of the main contributors to the deterioration of air quality are atmospheric aerosol particles. Compared to earlier manual approaches, MODIS, GIS, and remote sensing technologies have made AOT evaluation simpler. This article deals with the AOT assessment using GIS and Remote Sensing in Dhaka, Bangladesh in 2017. The necessary observations for the AOT assessment are made taking seasonal variation into account. To avoid overcorrecting the cloud cover, monsoon changes are not considered. When AOT is detected, the homogeneous surface condition is taken into account by the multispectral algorithm model. In essence, NASA Earth Observation (NEO) data are used for data validation, and Landsat 8+ OLI photos are used for AOT assessment. Compared to the pre- and post-monsoon seasons, it is discovered that winter has the largest concentration of AOT. This is caused by weather conditions such as clouds, rain, humidity, wind speed and pressure, temperature, etc.
Phuong D. et al. (2021)	Vietnam	Air Pollution Assessment Using RS and GIS in Ho Chi Minh City, Vietnam: A Case Study of Period 2015-2019 for SO_2 and NO_2 (Phuong et al., 2021)	Remote sensing photographs can be used to monitor and collect data on air pollution. This study uses data from the OMI of the AURA satellite of the A-Train group to estimate the level of NO_2 and SO_2 pollution in the atmosphere. Calculations and evaluations of NO_2 and SO_2 pollution levels will be performed. A map will be created showing the trend of NO_2 and SO_2 in Ho Chi Minh City from 2015 to 2019.

3 Limitations of GIS and Remote Sensing

There are several restrictions on RS-GIS science. The intricate interaction of different parts that make up an RS-GIS system makes it expensive. A fully qualified human workforce is necessary in addition to technology and software, but it is expensive to hire and train them (Limitations or Challenges of GIS, 2018). It merely offers some temporal, spectral, and geographical data. Remote sensing equipment frequently loses calibration, producing inaccurate remote sensing data. Finally, gathering, interpreting, and analyzing data from remote sensors may be expensive. Since the data from distant sensors are so useful, the cost is justified. It is necessary to use this technology to speed up the rapid monitoring of air pollution in metropolitan regions to receive accurate, prompt information on air pollution.

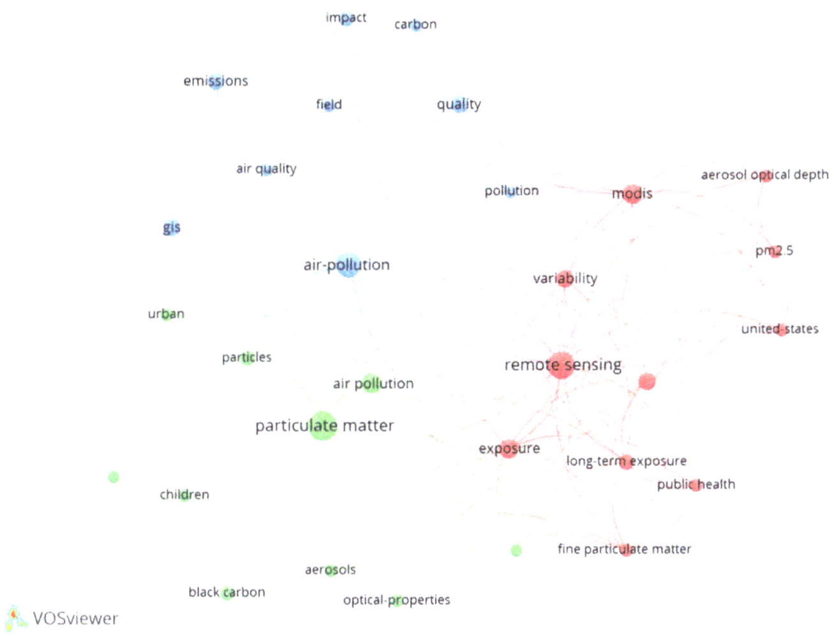

Figure 1: A bibliometric analysis and visualization of keyword co-occurrence in RS-GIS research on air pollution and its health effects.

References

Ahmad Jabatan Sains A, Pusat Perkhidmatan Akademik M, Hashim M, Md Hashim Pusat Pengajian Sosial N, dan Persekitaran P, Nizam Ayof Jabatan Sains M, Setyo Budi Jabatan Sains A (2006) The use of remote sensing and gis to estimate air quality index (AQI) over Peninsular Malaysia. https://api.semanticscholar.org/CorpusID:128157344

Akbari M, Zahmatkesh H, Eftekhari M (2021) A GIS-based system for real-time air pollution monitoring and alerting based on OGC sensors web enablement standards. Pollution 7(1):25–41 https://doi.org/10.22059/POLL.2020.296938.741

Al-Alola SS, Alkadi II, Alogayell HM, Mohamed SA, Ismail IY (2022) Air quality estimation using remote sensing and GIS-spatial technologies along Al-Shamal train pathway, Al-Qurayyat City in Saudi Arabia. Environ Sustain Indic 15:100184. https://doi.org/10.1016/j.indic.2022.100184

Applications of Remote Sensing in Air pollution (n.d.). https://grindgis.com/remote-sensing/9-applications-of-remote-sensing-in-air-pollution

Bai Y, Wu L, Qin K, Zhang Y, Shen Y, Zhou Y (2016) A geographically and temporally weighted regression model for ground-level PM2.5 estimation from satellite-derived 500 m resolution AOD. Remote Sens 8(3):262. https://doi.org/10.3390/rs8030262

Fuller R, Landrigan PJ, Balakrishnan K, Bathan G, Bose-O'Reilly S, Brauer M, Caravanos J, Chiles T, Cohen A, Corra L, Cropper M, Ferraro G, Hanna J, Hanrahan D, Hu H, Hunter D, Janata G, Kupka R, Lanphear B, Lichtveld M, Martin K, Mustapha A, Sanchez-Tiana E, Sandilya K, Schaefli L, Shaw J, Seddon J, Suk W, Yan C (2022) Pollution and health: a progress update. Lancet Planet Health 6(6):e535–e547. https://doi.org/10.1016/S2542-5196(22)00090-0

Geographic Information System (n.d.). https://education.nationalgeographic.org/resource/geographic-information-system-gis

GIS and Public Health at CDC|cdc.gov. (n.d.). https://www.cdc.gov/gis/index.htm. Accessed 22 Oct 2022

GIS Uses in Public Health (2019). https://grindgis.com/gis/gis-uses-in-public-health#:~:text=1.,waytomapthisdata

He J, Zha Y, Zhang J, Gao J (2014) Aerosol indices derived from MODIS data for indicating aerosol-induced air pollution. Remote Sens 6(2):1587–1604. https://doi.org/10.3390/rs6021587

Limitations or Challenges of GIS (2018). https://grindgis.com/remote-sensing/limitations-or-challenges-of-gis

Manisalidis I, Stavropoulou E, Stavropoulos A, Bezirtzoglou E (2020) Environmental and health impacts of air pollution: a review. Front Public Health 8:14. https://doi.org/10.3389/fpubh.2020.00014

McLafferty SL (2003) GIS and health care. Annu Rev Public Health 24(1):25–42. https://doi.org/10.1146/annurev.publhealth.24.012902.141012

Narashid RH, Wan Mohd WMN (2010) Air quality monitoring using remote sensing and GIS technologies. In: 2010 International Conference on Science and Social Research (CSSR 2010), pp 1186–1191. https://doi.org/10.1109/CSSR.2010.5773713

Nikolov H, Borisova D (2014) A framework for air quality monitoring based on free public data and open source tools (Michel U, Schulz K (eds)), p 92451P. https://doi.org/10.1117/12.2067328

Phuong DTK, Nhut MC, Tri ND (2021) Air pollution assessment using RS and GIS in Ho Chi Minh City, Vietnam: a case study of period 2015–2019 for SO_2 and NO_2. IOP Conf Ser: Earth Environ Sci 652(1):012004. https://doi.org/10.1088/1755-1315/652/1/012004

Sikder S, Rubaiyat Islam MA (2018) Aerosol optical thickness (AOT) assessment using GIS & remote sensing. SSRN Electron J 6:69–72. https://doi.org/10.2139/ssrn.3531752

Somvanshi SS, Vashisht A, Chandra U, Kaushik G (2019). Delhi air pollution modeling using remote sensing technique. In: Handbook of environmental materials management, pp 1–27. https://doi.org/10.1007/978-3-319-58538-3_174-1

Types of Remote Sensing: Technology Changing the World (n.d.). https://eos.com/blog/types-of-remote-sensing/

What is Remote Sensing? (n.d.). https://www.earthdata.nasa.gov/learn/backgrounders/remote-sensing

Analytical Methods for Oxygenated Polycyclic Aromatic Hydrocarbons (OPAHs) in Airborne Particulate: A Review

Noor E. Azam and Pratap Kumar Padhy

1 Introduction

Numerous toxic substances are present in $PM_{2.5}$ particles, for example extremely hazardous organic compound, heavy metal, virus and bacteria, which intensify their toxicity (Xu et al., 2018; Wang et al., 2016; Sun et al., 2018). Aside from various hazardous compounds in the atmosphere, PAHs and their oxygenated derivatives, known as OPAHs, are particularly dangerous in light of their unique carcinogenic properties. Occupational exposure to parent PAHs (PPAHs) is associated with an increased risk of lung, skin and bladder cancers (Huang et al., 2004; Ahn et al., 2006). In light of incomplete combustion of coal, diesel and gasoline fuel, there are a number of organic compounds known as PAHs. Zeng et al. (2018) reported, these compounds have fused aromatic rings (benzene) and are being related to a lot of traffic emissions.

Oxy-PAHs are polycyclic aromatic hydrocarbons containing carbonylic oxygen attached to one or more aromatic rings. In addition to being ubiquitous environmental pollutants, PAHs are found in a number of carcinogenic compounds. All types of organic matter produce PAHs as a byproduct of incomplete combustion (Nisbet and LaGoy, 1992). It was demonstrated that PAH byproducts contribute significantly in the formation of secondary organic aerosols in urban settings, highlighting the interaction between gas-phase PAHs and OH radicals (Lee et al., 2012). Moreover, PAHs can have the interaction with NO_x and O_3 in the atmosphere to produce or release oxy-PAHs (Albinet et al., 2007b). Lima et al. (2005) concluded that oxy-PAHs are commonly produced during complete combustion, especially under high-temperature conditions of 800 to 2000°C with 5-20% excess air.

N. E. Azam · P. K. Padhy (✉)
Department of Environmental Studies, Visva-Bharati, Santiniketan, Birbhum, West Bengal 731235, India
e-mail: pkpadhy@visva-bharati.ac.in

P. K. Padhy et al. (eds.), *Airborne Particulate Matter*,
https://doi.org/10.1007/978-3-031-84408-9_5

Within the oxy-PAH category, compounds include polycyclic aromatic quinones (PAQ), polycyclic aromatic ketones (PAK), hydroxylated PAHs (hydroxyPAHs), polycyclic aromatic carboxaldehyde, polycyclic aromatic lactones, polycyclic aromatic anhydrides, and polycyclic aromatic carboxylic acids. Due to the incorporation of oxygen atoms in their rings, they exhibit greater water solubility than their parent PAHs.

Certain oxy-PAHs have been found to be carcinogenic and mutagenic, potentially exhibiting greater toxicity than PAHs. (Albinet et al., 2007b; Yu, 2002). Investigations have been conducted in recent years into ambient oxy-PAH concentrations, their effects on human health, and their sources (Albinet et al., 2007b; Borgie et al., 2016; Bandowe et al., 2014; Tomaz et al., 2016). Oxy-PAH fractions assist in the induction of DNA adducts, supporting the possibility of genotoxic effects via pathways linked to parent PAHs, such as BaP (Du Four et al., 2004). Due to their increased bioavailability and direct mutagenic potential (Andersson and Achten, 2015), oxy-PAHs and nitro-PAHs may exhibit increasing toxicity than their unsubstituted parent PAHs. Compared to oxy-PAH derivatives, nitro-PAH derivatives have received less attention. Because of their greater oxidation state, nitro-PAHs are regarded as "dead-end products" of PAHs, which makes determining their potential toxicity challenging (Lundstedt et al., 2007). The magnitude of oxy-PAH concentrations is comparable to that of unsubstituted PAHs, approximately ten times higher than that of nitro-PAH concentrations (Huang et al., 2014; Albinet et al., 2007a; Wei et al., 2012; Wei et al., 2015; Harrison et al., 2016; Tomaz et al., 2016; de Oliveira Galvão et al., 2018; Bandowe and Nkansah, 2016).

Oxygenated PAHs are shown to be important in toxicity (Ohyama et al., 2007), and as a result, more research is being done on their chemical makeup, prevalence, destiny and behaviour. There have been a number of notable contributions to this field (Lee et al., 2018; Nežiková et al., 2021; Elzein et al., 2019; Zhuo et al., 2017; Dos et al., 2020; Shen et al., 2016; Zhang et al., 2018a; Hu et al., 2019; Lei et al., 2021; Zhang et al., 2018b). Oxygenated PAHs are introduced in this article in terms of: (i) their chemical and physical characteristics, (ii) their primary sources and pathways into the atmosphere, and (iii) the effects they have on health. The paper highlights the progress made in analyzing the chemical composition of oxygenated PAHs absorbed into particulate matter (PM). An emphasis is on: (i) collection of samples and artifacts, (ii) preparation of samples, such as extracting, cleaning, preconcentrating and fractionating, and (iii) separation and analysis of samples.

2 Physical-Chemical Properties

In the past, oxygenated PAHs have been investigated experimentally in limited numbers. Using SciFinder Scholar's ACD Solaris Estimation algorithm, theoretical estimations of oxygenated PAHs can provide valuable insights into their toxicological and environmental behaviours when experimental data is lacking (Walgraeve et al., 2010). Oxygenated PAHs exhibit a higher vapour pressure and molecular weight

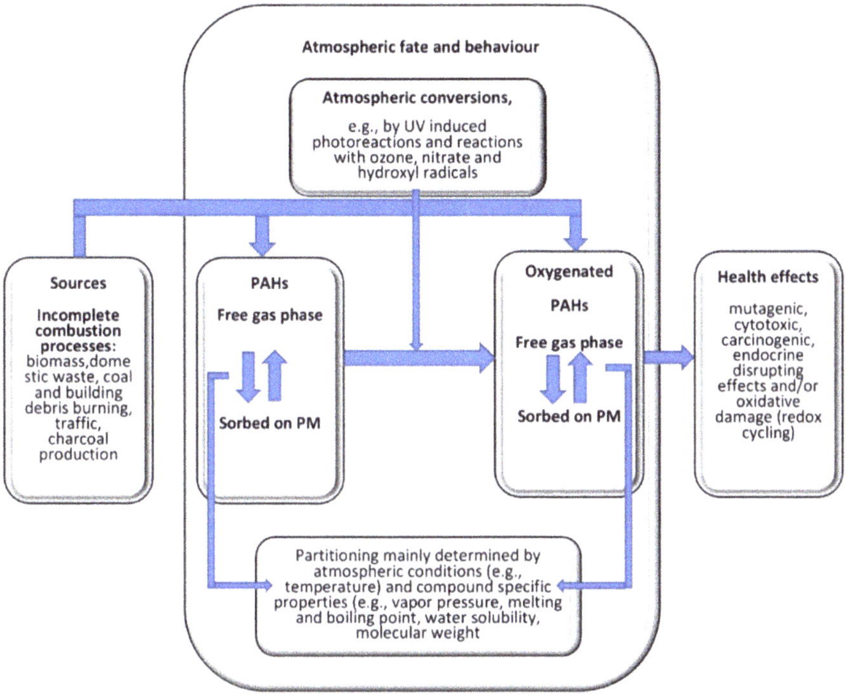

Figure 1: Illustration of the main sources and pathways of oxygenated PAHs and their effects on human health. (Source: Walgraeve et al., 2010)

than their parent PAHs (Goldfarb and Suuberg, 2008). The vapour pressure plays an important role in determining how organic compounds partition from the gaseous phase to particulates (Fig. 1). Under ambient conditions, oxygenated PAHs exhibit a higher propensity to associate with particulates than the parent PAHs (Vione et al., 2004), as their vapour pressures are well below 10 Pa. Section 3 discusses this matter in detail.

3 Sources, Fate and Atmospheric Pathways

Oxygenated PAHs differ from PAHs due to their dual origin: they are directly emitted and are also created in the troposphere through PAH conversion (Albinet et al., 2007a; Vu-Duc et al., 2007; Lemieux et al., 2004) or other pioneer molecules (Schauer et al., 2004; Webb et al., 2006; Feilberg et al., 2002; Phousongphouang and Arey, 2003; Morel et al., 2006). Gaseous and particulate matter-associated oxygenated PAHs can be directly produced by incomplete combustion processes involving a multitude of materials. Household waste and biomass (Gullett et al., 2003; Sidhu et al., 2005) are primarily responsible for these emissions.

Oxidated PAHs are produced through photochemical reactions and interactions between different radicals, such as nitrate radicals, hydroxyl radicals and ozone. Due to the fact that PAHs occur in both gaseous and particulate forms, they undergo transformation processes in both phases. Gas-phase reactions are primarily influenced by hydroxyl radicals (in daytime), and nitrate radicals (at night) (Vione et al., 2004), while oxygen and photolysis serve as secondary factors. PAHs associated with particles having greater than five rings in gaseous solutions are a smaller amount likely to react with reactive radicals. Despite their slower rate of transformation than gas-phase reactions, it is still possible to transform these compounds in a variety of ways. Direct photolysis has a major effect on these compounds when photosensitizers like quinones or aromatic carbonyls are present. PAHs absorb sunlight through extensive electron delocalization, leading to photooxidation under atmospheric conditions (Vione et al., 2006).

When it is in gaseous form, primary reactions of PAHs with OH radicals, NO_3 radicals, or O_3 produce energy-rich adducts (Atkinson and Arey, 2007). Oxy-PAH levels in the atmosphere are influenced by several factors, including compounds with three rings, such as anthracene, or four rings, like pyrene. As intermediate products in complex reactions that ultimately produce quinones (Keyte et al., 2013), these hydroxylated compounds serve as intermediate products. Furthermore, their study demonstrated that oxygenated PAHs, like PAH-NO_3, may be more harmful to the environment than other structurally altered polycyclic aromatic hydrocarbons. It is oxy-PAHs present in the environment that is caused by a group of compounds with various molecular structures, such as polycyclic aromatic ketones and polycyclic aromatic quinones. Several studies highlight the importance of hydroxy-PAHs in reactions, the slower production of polycyclic aromatic lactones, and the importance of PAQs and PAKs as PAH degradation products. According to this evidence, PAQs, PAKs or the parent PAHs are found at higher concentrations in the environment than hydroxy-PAHs or carboxy-PAHs (Barrado et al., 2012; Kamiya et al., 2005; Kishikawa et al., 2004; Bandowe et al., 2014).

Oxygenated PAHs behave differently in the atmosphere depending on the distribution between particles and gaseous phases. As a general guideline proposed by Howard and Meylan (vapour pressures below 1.3×10^{-6} Pa), substances with low vapour pressures exhibit a substantial affinity for particulate matter, resulting in minimal presence in the free gas phase. Oxygenated PAHs are partitioned between these phases based on limited experimental evidence. Under equilibrium conditions, oxygenated PAHs are strongly associated with particulate phases based on their physicochemical characteristics, including vapour pressure and molecular weight.

It is important to know that air serves as the primary pathway for releasing all polycyclic aromatic compounds (PACs), and how these PACs split into gas and particulate matter is determined by their physical characteristics. Researchers in Grenoble observed how vapour pressure and molecular weight affected the gas-particle partitioning of PACs over a year (Tomaz et al., 2016). They found that PACs with lower molecular weights (less than 202 g/mol) stayed mostly in the gas phase, while PACs with higher molecular weights (more than 252 g/mol) dominated the particle phase. Another study in the crowded Pearl River Delta area of China examined how sunlight, wind speed, O_3, NO_X, and SO_2 levels influenced the formation of PAKs and PAQs (Huang et al., 2014). They discovered a significant link between sunlight, wind speed and SO_2, indicating that these factors mattered. They also noticed that PAQ levels rose sharply from October to March, showing a clear seasonal effect. This rise in emissions during this period could be linked to winter and home heating practices (Bandowe et al., 2014; Tomaz et al., 2016).

Diesel particulate matter exhibited the highest levels of PAK and PAQ components, measured at parts per million (ppm) levels. Notably, Nocun and Schantz (2013) found that BAQ or 7,12-Benz[a]anthracene quinone, had a concentration of 5 mg/kg, higher than other components. Oxy-PAHs deserve considerable research due to their concentrations comparable to PAHs, and sometimes even higher than nitro-PAHs, especially concerning PAKs and PAQs.

4 Human Health Effects

4.1 Exposure Routes

A majority of particulate matter is deposited in the respiratory tract, which affects the pulmonary alveoli and the tracheobronchial tree. A particulate matter is classified as ultrafine ($PM_{0.1}$), fine ($PM_{2.5}$), and coarse ($PM_{2.5\text{-}10}$). The diameter of fine particles is less than 2.5 micrometers, particles with a diameter between 2.5 and 10 micrometers are considered coarse. Although $PM_{2.5\text{-}10}$ is usually deposited and cleared in the upper respiratory tract, $PM_{0.1}$ and $PM_{2.5}$ could reach the alveoli, where they are cleared at a slower rate. Further, the specific surface area of these minute particles is large and capable of absorbing various substances, including oxy-PAHs (Oh et al., 2011; Kim and Hu, 1998). According to a study by Boublil et al. (2013), which examined three different particle sizes, bronchial epithelial cells serve as the first targets of particulate matter exposure. As soon as oxygen-PAHs are inhaled or ingested, they can react with DNA and proteins in the human body, potentially causing genotoxicity.

4.2 Health Effects

When it comes to oxygenated PAHs and their health effects, they are frequently taken into account as more hazardous in comparison to their parent PAHs due to their inherent mutagenic potential, while parent PAHs require enzyme activation (Yu, 2002). Various factors contribute to their toxicity, but the mechanisms remain unclear (Cho et al., 2005; Lundstedt et al., 2007; Lampi et al., 2006). Based on specific end-points studied, oxidized PAHs can differ in their toxicological potency. As a group of compounds, oxygenated hydrocarbons are involved in cardiovascular disease complaints, as stated in the study by Metzger et al. (2004). Furthermore, oxygenated

PAHs have been shown to be endocrine-disrupting agents, exhibiting both estrogenic and antiestrogenic effects, in addition to increased cytotoxicity compared to their parent PAHs (Kishikawa et al., 2004). As compared to methoxyphenols and oxygenated PAHs, wood smoke particulate matter (that was extracted at temperatures ranging from 100 to 150 °C) exhibited increased cytotoxicity (Kubatova et al., 2004).

There is a proposed role for quinones in PM toxicity (Li et al., 2003; Taguchi et al., 2007, 2008). Quinones and their reduced forms, including semiquinones and hydroquinones, have drawn considerable attention (Taguchi et al., 2007). In consequence of redox cycling with their respective semiquinones, these compounds can produce reactive oxygen species. Various reactive oxygen species, including superoxide anion radicals (O^{2-}) and hydroxyl radicals (HO^{-}), can induce severe oxidative stress. These species can cause lipid, protein and DNA oxidation. According to Chung et al. (2007), oxidative stress occurs when free radicals and oxidants damage cells, leading to inflammation and apoptosis. As reported by Kubatova et al. (2006) oxygenated PAHs are associated with pulmonary diseases and oxidative DNA damage, including 1,8-dihydroxy and anthracene-9,10-dione. The toxicological significance of oxygenated PAHs underscores the importance of measuring them within atmospheric particulate matter.

DNA single-strand breaks and cell division frequency can be investigated using a variety of assays, such as Comet and CBMN (Fairbairn et al., 1995; Jha, 2008; Fenech, 1993; Fenech et al., 1999). In BEAS-2B cells, crude extract (CE) significantly increased the development of micronuclei and damaged DNA, suggesting that it is genotoxic. Topinka et al. (1998) reported that the aliphatic, aromatic and moderately polar portions of CE generated micronuclei that were comparable to each other and seriously damaged DNA. Despite the fact that precise information on their composition is lacking, the genotoxic effects of these fractions may be partially attributed to oxy-PAHs in this study. Acenaphthenequinone and BAQ are genotoxic to Japanese medaka (Oryzias latipes) embryos, according to Dasgupta et al. (2014). In Oryzias latipes embryos, a dose-dependent response to DNA breakage was noted.

Sharma et al. (2012) pointed out that evidence has demonstrated a substantial correlation between H2AX phosphorylation and DNA damage, specifically double-strand breaks. In less than 24 hours, exposure to low concentrations of 0.01 mg/mL of oxy-PAH phosphorylated the histones of H2AX. This study suggests that oxy-PAHs may be genotoxic since they are 100 times more harmful to DNA than PAHs. To validate this genotoxicity and learn more about how they interact with cells, more research is required.

By interacting with nucleotides at various sites, activated oxy-PAHs can form DNA adducts. Umbuzeiro et al. (2008) utilized a highly sensitive radioactive 32P-post labeling method to examine the effects of exposure, DNA adducts are formed to fractions of urban particulate matter (UPM) containing oxy-PAHs at two distinct locations in São Paulo, Brazil.

4.3 *Other* in vitro *Endpoints Implicated in Potential Carcinogenic Risk of Oxy-PAHs*

The genotoxic and mutagenic potential of a drug is an important determinant of carcinogenic risk when assessing its mode-of-action (MOA). Even if this essential event has only partially been confirmed, the ability of oxyPAHs to alter DNA sequences within target organs, where tumors are suspected, represents a pivotal and measurable endpoint. By facilitating the bioactivation of pro-genotoxic compounds, certain enzymes involved in xenobiotic metabolism can indirectly and non-specifically induce mutagenic effects. Since numerous carcinogenic and mutagenic substances require prior cellular biotransformation to manifest their toxicity (Kopponen et al., 1994; Topinka et al., 1998), enzyme induction such as CYP1A1 serves as another endpoint in assessing carcinogenic risk. The oxidation of PAHs can in-turn, increase intracellular oxy-PAH concentrations through the induction of this enzyme.

Furthermore, Wislocki et al. (1976) studied Chinese hamster lung cells, while Durant et al. (1996) tested human B lymphoblastoid cells (h1A1v2) for mutations at a tyrosine kinase gene locus. Even though oxy-PAHs have been tested for potential toxic effects including enzyme induction/inhibition, genotoxicity, mutagenicity and tumor promotion, the results of these assessments remain inconclusive regarding their precise toxic effects.

During the initial phase of the air study, various fractions of UPM sample are evaluated for their global mutagenicity. *S. thyphimurium* strains TA98 and YG1041 were tested for resistance using the Ames reverse mutation assay by Umbuzeiro et al. (2008). According to their findings, organic extracts containing UPM, including non-polar PAHs listed by the US EPA, showed minimal mutagenic activity (Umbuzeiro et al., 2008). Mutagenic response was significant for UPM extracts containing oxyPAH congeners, consistent with the outcomes observed for individual compounds. In light of these results, it is plausible that these polar fractions include the parent compound BANTone, along with several other target oxyPAHs. They suggest that these polar fractions have mutagenic properties attributed to undiscovered oxy-PAHs.

During their study, Durant et al. (1996) selected 67 specific individual compounds to identify persistent aromatic compounds in urban aerosols. The researchers tested how these chemicals could cause mutations in human B cells (h1A1v2) that had tyrosine kinase markers. They saw a dose-response curve for phenalenone, CPPO, BANTone, BPO and anthanthrene quinone at the doses they used.

5 Analytical Sequence for Oxygenated PAHs Determination on the Samples of Particulate Matter

5.1 Sampling Artefacts for Oxygenated PAHs

Nežiková et al. (2021) used high-volume samplers to get samples. They got the vapour phase with foam discs and the particle phase with quartz filters and cleaned the discs for eight hours with acetone and DCM before using them. They put each sample in a ziplock bag, covered it with aluminum foil, and stored it at –18°C until analysis.

Elzein et al. (2019) used a big sampler with high flow rates of 1.33 m^3 per min to get $PM_{2.5}$ filter samples from 8 m high. They heated the 20.3 × 25.4 cm quartz filters at 550°C for 5 hours to remove any organic stuff and vacuum-sealed the filters and kept them at –20°C until analysis, with plastic bags and aluminum foil for protection.

Xie et al. (2014) studied SVOCs in gas and particle phases. They used a sampler and two sampling trains with cyclones that ran at 92 liters per minute and had a 2.5 mm cut. They used different filters like QFFs, PUF/XAD-4/PUF, and XAD-7 resin to filter the air. The main sample train got all the compounds they wanted, while some filters checked how the gas stuck to the media. They used many PUF/XAD-4/PUF sandwiches to see the breakthrough of each compound over eight days. Liu et al. (2006) got particulate matter for organic chemical analysis with glass, quartz, or Teflon fiber filters and used low-volume or high-volume samplers for 2 to 48 hours. The high-volume samplers had flow rates from 25 to 100 m/h, while the low-volume samplers had flow rates from 1 to 1.8 m^3/h. Ozone, nitrate and hydroxyl radicals in the air could react with PAHs and oxygenated PAHs during sampling, making sampling artifacts. They used ozone denuders to take out ozone from the air before getting particles on filters to lower the risk of wrong concentrations and health effects.

Liu et al. (2006) looked at how manganese oxide and ozone denuders changed oxygenated PAH samples from low-volume sampling and split 2- to 3-ring and 4- to 5-ring oxygenated polycyclic aromatic hydrocarbons because they acted differently. Oxygenated PAHs were higher without an ozone denuder, like 11H-benzo[de]anthracene-7-one, 7H-benzo[de]anthracene-11-one, benzo[a]anthracene-7,12-dione and 11H-benzo[b]fluorene-11-one. These cases were more likely from ozone making than breaking. When they used an ozone denuder, the amount of 4H-cyclopenta[cd]pyrene-3-one went down by about 10%, showing a bigger chance of ozonation. The amount of 6H-benzo[cd]pyrene-6-one, however, stayed the same.

When quartz fiber filters were used for high-volume sampling at 20 to 30°C, anthracene-9,10-dione was reduced by 40 to 50% (Leoz-Garziandia et al., 2000). Moreover, when compared with non-denuded samples, denuded samples showed an increase in the average concentrations of two- to three-ring oxygenated PAHs, such as naphthalene-1,4-dione, 4H-cyclopenta[def]phenanthrene-4-one, anthracene-9,10-dione, and 9H-fluorene-9-one. In these oxygenated PAHs, catalytic formation dominates ozone degradation at the denuder surface, based on this observation. Therefore, gaseous oxygenated PAHs cannot be analyzed by a MnO_2

(manganese dioxide) ozone denuder. Also, oxygenated PAHs are more likely to be lost through vaporization during sampling.

As reported by Dos et al. (2020), this study used a high-volume sampler to gather samples of suspended particles of $PM_{2.5}$. The sampling process involved the use of two types of filters: quartz filters measuring 20.3 × 25.4 cm with a thickness of 47 mm and polyurethane foam (PUF) measuring 21 × 26 × 0.3 cm. The researchers removed water and organic stuff from the quartz filters by heating to 400°C for four hours. They cleaned the PUFs in an ultrasonic bath with a gentle soap solution for half an hour and rinsed and dried them for two and a half hours at 100°C. They kept the PUFs and quartz filters in a desiccator until they needed them and got the gas phase with the PUF and the particle phase with the quartz filter. They covered these filters with aluminum foil and stored them at −18°C until they looked at them and got the $PM_{2.5}$ by weighing the quartz filters before and after sampling.

5.2 *Sample Preparation, Extraction, Clean-Up, Analysis, Separation and Detection*

In Elzein et al. (2019) study, the extraction process involved using an accelerated solvent extractor to process all the collected samples. With a hole puncher, 1/16th of each filter, covering 25.7 cm^2, was cut prior to extraction. The filters were treated with deuterated-OPAHs, specifically 9,10-Anthraquinone-D8 and 9-Fluorenone-D8 at a concentration of 400 nanograms (40 μl at 10 ng μl^{-1} in Acetonitrile). In order to extract the samples, they were punched, divided into smaller pieces and placed within five milliliter stainless steel extraction cells. With a rinse volume of 60% and a purge time of 60 sec, three consecutive 5-minute acetonitrile extraction cycles were performed at 120°C and 1500 psi. Following extraction, nitrogen was gently streamed to evaporate the extracts (V=20ml) to approximately 6 mL before further processing. Solid-phase extraction (SPE) was employed to purify all samples and blanks in order to preserve the GC inlet liner. One milliliter of the sample solution was evaporated to 1.5 milliliters at room temperature (20°C) under a mild nitrogen stream, and the milliliters were then transferred to amber autosampler vials. The concentrated samples were kept at 4°C until analysis. Elzein et al. (2019) determined compound concentrations based on how well the compounds were recovered, which ranged between 85 and 96%.

Lee et al. (2018) cut the filter into two pieces with acetone-cleaned scissors and put 25 milliliters of deuterated PAHs with 20 milligrams per milliliter in each piece. They did not have isotopically labeled standards, hence used deuterated compounds like acenaphthene-d10, naphthalene-d8, chrysene-d12, phenanthrene-d10, perylene-d12, and 1,4-dichlorobenzene-d4 instead of oxy-PAH standards. However, using deuterated PAHs for analyzing oxy-PAHs might have some limitations. To extract from the filter, they used a mixture of methanol (3 parts) and dichloromethane (DCM), and then stored these extracts in an amber vial for future use. For an additional thirty minutes, a second extraction procedure using fresh DCM solvents and methanol (3:1, v/v) was carried out. In order to decrease the extract solution's volume, a nitrogen gas stream was used to evaporate it. In order to purify the extract solution, the syringe filter utilized was 0.45 mm, and a nitrogen concentrator was used to concentrate it to 250 ml. A Hewlett Packard 7890A GC with a mass selectivity detector for 5975 was used to conduct the GC/MS study. To conduct ion monitoring/scanning using electron impact, HP-5MS capillary columns measuring 30 m by 0.25 mm by 0.25 m were used. A GC oven was preheated to 60°C for 1 min, and then for 5 min it was raised to 310°C at a pace of 6°C per minute. Using internal standardization, oxy-PAHs were quantified. All oxy-PAH compounds were calibrated using seven calibration standard solutions ranging in concentration from 0.1 ng/mL to 100 ng/ml. Lee et al. (2018) reported excellent calibration results ($r^2 > 0.998$).

To extract PUF samples, Zhuo et al. (2017) used 150 mL of hexane and acetone (at a 1:1 ratio) for 8 hours. The researchers used 25 mL of hexane and acetone (1:1) for microwave-accelerated extraction (MAE). They heated to 110°C for 20 min and got the extracts, cleaned, and put them on a silica/alumina gel column after making them 1.0 mL with a rotary evaporator. They studied oxy-polycyclic aromatic hydrocarbons (OPAHs) in environmental samples with different methods. For example, Zhuo et al. (2017) washed the column with 20 mL of hexane and 70 mL of a hexane/dichloromethane (1:1) mix and made the eluate about 1 mL and added deuterated internal standards. They looked at Oxy-PAHs with gas chromatography-mass spectrometry (GC-MS) in Negative Chemical Ionization (NCI) mode. They used an HP-5MS capillary column with helium and methane and raised the temperature from 60 to 150°C by 15°C per min, then kept it at 300°C for 15 min. They found oxy-PAHs by their specific ions and retention times of standards like 9FO, BZO, ATQ and BaAQ.

Shen et al. (2016) got gaseous pollutants from PUFs with a Soxhlet extraction method with 150 mL of dichloromethane after 8 hours and took out particles from the filters with acetone and hexane (20 mL) and a 1200 W microwave. They cleaned the extracted material with a silica/alumina gel column and then used an HP-5MS capillary column for GC-MS analysis after making the eluate 1 mL and adding deuterated internal standards. Dos et al. (2020) used cold-fiber solid-phase microextraction (CF-SPME) in their method and used acetonitrile, ultrapure water and quartz filter discs in a 200 mL vial. They shook it in an ultrasonic bath with PDMS/DVB fibers cooled by liquid nitrogen and heated it to 270°C in the gas chromatograph injector for analysis using an Agilent HP-5MS capillary column. Zhang et al. (2018a, b) mixed samples from three circular filters and split them with dichloromethane for eight hours through an alumina/silica gel column. They concentrated the eluates and analyzed them with gas chromatography-mass spectrometry (GC-MS) with internal standards and used high-purity nitrogen and GC-MS/MS on an HP-5MS column with selective ion monitoring (SIM) and negative chemical ionization (NCI) to study OPAHs. Hu et al. (2019) added anthraquinone-D8, an OPAH surrogate, before extracting to help examine the target molecules and used ultrasonic extraction cycles and Soxhlet extraction with dichloromethane for PUF samples. These methods helped them study OPAHs in various environmental samples. They got the wastewater, almost concentrated it with a gentle nitrogen stream, and dissolved it in 300 microliters of isooctane.

They used a gas chromatography/mass spectrometry (GC/MS) device with DB-5MS columns (30 m long, 0.25 mm wide, and 0.25 mm thick, J&W Scientific). Hu et al. (2019) measured ten OPAHs with the negative chemical ionization method and used a selected ion monitoring (SIM) setup for the analysis. They measured three OPAHs' concentrations in this study with a thermal desorption/mass spectrometry (TD-GC/MS) system in the injection port. Before samples were loaded, injector ports were adjusted to 275°C and 50°C, and the GC oven was kept at a temperature of about 30°C. Based on a study by Lei et al. (2021), benzophenone-D10 and anthraquinone-D8 were presented as internal standards for OPAH measurement.

Dichloromethane (DCM) and n-hexane, both of which had purity levels appropriate for pesticide detection, were employed (Zhang et al., 2018a, b). On specific sampling days when concentrations were low, composite filter samples were extracted in groups to guarantee enough quantities of the target chemicals for measurement. Prior to the extraction procedure, surrogate standards were applied in order to quantify OPAHs and 150 milliliters of DCM were used to extract the composite filter samples over the course of 8 hours. The identical procedure, as previously described, was used to extract each PUF plug separately, and the resulting extracts were combined. Through the use of an evaporator with a thermostatic bath set 30°C, the extracts were concentrated to around 1 milliliter. Alumina and silica gel columns were used for purification. Cotton wool served as the basis, and 12 cm of silica, 12 cm of alumina, and 1 cm of anhydrous sodium sulfate were used to make the top. After 20 mL of n-hexane was used to elute the column, 70 mL of DCM and n-hexane were used to collect the sample. Following further concentration of the extract, 10 mL of n-hexane was added, and the extract was evaporated down to 1-2 mL using a rotary evaporator. They were able to attain an additional concentration of 1 milliliter in a thermostatic bath with a nitrogen stream that was 99.999% pure at 30°C. Using anthraquinone-d8, injections of 100 ng of perdeuterated internal standards were made into a gas chromatograph-mass spectrometer (GC-MS). The OPAHs were examined using an HP-5MS column in SIM mode with negative chemical ionization. The researchers set the GC oven to start at 60°C for 1 min and then go up by 15°C per min until it got to 150°C. Then it went up by 5°C per min until it reached 300°C, where it stayed for 15 min. This procedure followed Zhang and his team's (2018a, b) method.

Nežiková et al. (2021) used a high-temperature Soxhlet extraction method with DCM to get compounds and checked the samples with surrogate recovery criteria. After separating the samples with a silica column with 1 g of dry sodium sulfate and 5 g of silica gel they washed the column with 50 milliliters of DCM and 10 milliliters of *n*-hexane and split the samples into different vials and let them evaporate with a gentle nitrogen stream until they were about 150 milliliters. The PCB121, an OPAH standard, was used to study the syringes and the samples were analyzed with gas chromatography and mass spectrometry. The procedure included using the internal standard (PCB121) together with a deuterated surrogate standard. With a high recovery rate of 70 to 105%, the researchers were able to extract native OPAH molecules from spiked polyurethane foam samples. Chemical ionization tandem mass spectrometers and gas chromatography 7890 A were employed in the investigation. The gas chromatograph was equipped with a 30 meter by 0.25 millimeter by 0.25 micrometer Rxi-5Sil MS column. Using a dry source and a 1 milliliter injection of splitless helium at 270°C, several reactions were observed. Helium was employed as a carrier gas, flowing at a constant 1.5 milliliters per min. The oven temperature was raised by 40°C per min until 180°C was attained after it was first set at 90°C for 1 min. The temperature was maintained at 320°C for 6 min at a rate of 5°C per min. The ion source was set at 150°C, while the transfer line ran at 290°C in this investigation (Nežiková et al., 2021).

6 Conclusions

Because oxygenated PAHs can be harmful and cause mutations, it is crucial to keep track of where they are, what happens to them, and how they behave. Sunlight and air releases can change these compounds and make them very small in the air. They stick to little bits and sprays because they hardly evaporate, especially when it is not hot. This is bad because they can enter into our lungs and blood and hurt us by making a lot of oxidation. We need to study their physical and chemical features to understand how these oxygenated PAHs come from different places, both straight and twisted, and how they affect our bodies and environment. It is hard to measure these oxygenated PAHs in the fine dust in the air. First, it is crucial to make sure the samples collected, whether they are small or large in volume, do not have any mistakes or changes that could have happened during collection. A promising avenue of research is to examine technological advances like ozone denuders. A number of traditional extraction methods, including Soxhlet and ultrasonic extraction, have also been explored internationally in various environmental analysis fields. Novel solvent extraction techniques, such as pressured liquid extraction, microwave extraction, and supercritical fluid extraction are currently used to study oxygenated PAHs. As part of the separation phase, gas chromatography is commonly used in conjunction with electron impact ionization. Nonetheless, ongoing research in this field is exploring alternative approaches like: (i) liquid chromatography (LC)-based separations, (ii) multidimensional chromatography (e.g., GC×GC, LC-GC) combined with mass spectrometry, and (iii) the integration of online sample preparation with separation/detection processes.

Due to the fact that oxygenated PAHs are found in trace amounts in the atmosphere and are prevalent in intricate environments, it is imperative that sensitive and specific analytical techniques are used in order to handle a large volume of samples. For monitoring purposes, numerous samples must be analyzed. A number of key challenges await future research: (i) gaining a better understanding of oxygenated PAHs' physical and chemical characteristics, (ii) creating precise analytical techniques to detect and measure PAHs that are oxygenated in the atmosphere, including method validation, and (iii) developing comprehensive monitoring programs to understand the factors that affect oxygenated PAH occurrence. Consideration must be given to their effects on human health, taking into account specific sources, atmospheric mechanisms and geographic regions.

References

Ahn YS, Park RM, Stayner L, Kang SK, Jang JK (2006) Cancer morbidity in iron and steel workers in Korea. Am J Ind Med 49(8):647–657

Albinet A, Leoz-Garziandia E, Budzinski H, Villenave E (2007a) Sampling precautions for the measurement of nitrated polycyclic aromatic hydrocarbons in ambient air. Atmos Environ 41(23):4988–4994

Albinet A, Leoz-Garziandia E, Budzinski H, Villenave E (2007b) Polycyclic aromatic hydrocarbons (PAHs), nitrated PAHs and oxygenated PAHs in ambient air of the Marseilles area (South of France): concentrations and sources. Sci Total Environ 384(1–3):280–292

Andersson JT, Achten C (2015) Time to say goodbye to the 16 EPA PAHs? Toward an up-to-date use of PACs for environmental purposes. Polycycl Aromat Compd 35(2–4):330–354

Atkinson R, Arey J (2007) Mechanisms of the gas-phase reactions of aromatic hydrocarbons and PAHs with OH and NO_3 radicals. Polycycl Aromat Compd 27(1):15–40

Bandowe BAM, Nkansah MA (2016) Occurrence, distribution and health risk from polycyclic aromatic compounds (PAHs, oxygenated-PAHs and azaarenes) in street dust from a major West African Metropolis. Sci Total Environ 553:439–449

Bandowe BAM, Meusel H, Huang RJ, Ho K, Cao J, Hoffmann T, Wilcke W (2014) $PM_{2.5}$-bound oxygenated PAHs, nitro-PAHs and parent-PAHs from the atmosphere of a Chinese megacity: Seasonal variation, sources and cancer risk assessment. Sci Total Environ 473:77–87

Barrado AI, García S, Barrado E, Pérez RM (2012) $PM_{2.5}$-bound PAHs and hydroxy-PAHs in atmospheric aerosol samples: correlations with season and with physical and chemical factors. Atmos Environ 49:224–232

Borgie M, Ledoux F, Dagher Z, Verdin A, Cazier F, Courcot L, Shirali P, Greige-Gerges H, Courcot D (2016) Chemical characteristics of $PM_{2.5–0.3}$ and $PM_{0.3}$ and consequence of a dust storm episode at an urban site in Lebanon. Atmos Res 180:274–286

Boublil L, Assémat E, Borot MC, Boland S, Martinon L, Sciare J, Baeza-Squiban A (2013) Development of a repeated exposure protocol of human bronchial epithelium in vitro to study the long-term effects of atmospheric particles. Toxicol in Vitro 27(2):533–542

Cho AK, Sioutas C, Miguel AH, Kumagai Y, Schmitz DA, Singh M, Eiguren-Fernandez A, Froines JR (2005) Redox activity of airborne particulate matter at different sites in the Los Angeles Basin. Environ Res 99(1):40–47

Chung SW, Chung HY, Toriba A, Kameda T, Tang N, Kizu R, Hayakawa K (2007) An environmental quinoid polycyclic aromatic hydrocarbon, acenaphthenequinone, modulates cyclooxygenase-2 expression through reactive oxygen species generation and nuclear factor kappa B activation in A549 cells. Toxicol Sci 95(2):348–355

Dasgupta S, Cao A, Mauer B, Yan B, Uno S, McElroy A (2014) Genotoxicity of oxy-PAHs to Japanese medaka (*Oryzias latipes*) embryos assessed using the comet assay. Environ Sci Pollut Res 21(24):13867–13876

de Oliveira Galvao MF, de Oliveira AN, Ferreira PA, Caumo S, de Castro VP, Artaxo P, de Souza HS, Roubicek DA, de Medeiros SRB (2018) Biomass burning particles in the Brazilian Amazon region: mutagenic effects of nitro and oxy-PAHs and assessment of health risks. Environ Pollut 233:960–970

Dos Santos RR, de Lourdes Cardeal Z, Menezes HC (2020) Phase distribution of polycyclic aromatic hydrocarbons and their oxygenated and nitrated derivatives in the ambient air of a Brazilian urban area. Chemosphere 250:126223

Du Four VA, Van Larebeke N, Janssen CR (2004) Genotoxic and mutagenic activity of environmental air samples in Flanders, Belgium. Mutat Res/gen Toxicol Environ Mutagen 558(1–2):155–167

Durant JL, Busby WF Jr, Lafleur AL, Penman BW, Crespi CL (1996) Human cell mutagenicity of oxygenated, nitrated and unsubstituted polycyclic aromatic hydrocarbons associated with urban aerosols. Mutat Res/gen Toxicol 371(3–4):123–157

Elzein A, Dunmore RE, Ward MW, Hamilton JF, Lewis AC (2019) Variability of polycyclic aromatic hydrocarbons and their oxidative derivatives in wintertime Beijing, China. Atmos Chem Phys 19(13):8741–8758

Fairbairn DW, Olive PL, O'Neill KL (1995) The comet assay: a comprehensive review. Mutat Res/rev Gen Toxicol 339(1):37–59

Feilberg A, Ohura T, Nielsen T, Poulsen MWB, Amagai T (2002) Occurrence and photostability of 3-nitrobenzanthrone associated with atmospheric particles. Atmos Environ 36(22):3591–3600

Fenech M (1993) The cytokinesis-block micronucleus technique and its application to genotoxicity studies in human populations. Environ Health Perspect 101(suppl 3):101–107

Fenech M, Crott J, Turner J, Brown S (1999) Necrosis, apoptosis, cytostasis and DNA damage in human lymphocytes measured simultaneously within the cytokinesis-block micronucleus assay: description of the method and results for hydrogen peroxide. Mutagenesis 14(6):605–612

Goldfarb JL, Suuberg EM (2008) Vapor pressures and thermodynamics of oxygen-containing polycyclic aromatic hydrocarbons measured using knudsen effusion. Environ Toxicol Chem: Int J 27(6):1244–1249

Gullett BK, Touati A, Hays MD (2003) PCDD/F, PCB, HxCBz, PAH, and PM emission factors for fireplace and woodstove combustion in the San Francisco Bay region. Environ Sci Technol 37(9):1758–1765

Harrison RM, Alam MS, Dang J, Ismail IM, Basahi J, Alghamdi MA, Hassan IA, Khoder M (2016) Relationship of polycyclic aromatic hydrocarbons with oxy (quinone) and nitro derivatives during air mass transport. Sci Total Environ 572:1175–1183

Hu H, Tian M, Zhang L, Yang F, Peng C, Chen Y, Shi G, Yao X, Jiang C, Wang J (2019) Sources and gas-particle partitioning of atmospheric parent, oxygenated, and nitrated polycyclic aromatic hydrocarbons in a humid city in southwest China. Atmos Environ 206:1–10

Huang B, Liu M, Bi X, Chaemfa C, Ren Z, Wang X, Sheng G, Fu J (2014) Phase distribution, sources and risk assessment of PAHs, NPAHs and OPAHs in a rural site of Pearl River Delta region, China. Atmos Pollut Res 5(2):210–218

Huang W, Grainger J, Patterson DG, Turner WE, Caudill SP, Needham LL, Pirkle JL, Sampson EJ (2004) Comparison of 1-hydroxypyrene exposure in the US population with that in occupational exposure studies. Int Arch Occup Environ Health 77(7):491–498

Jha AN (2008) Ecotoxicological applications and significance of the comet assay. Mutagenesis 23(3):207–221

Kamiya M, Toriba A, Onoda Y, Kizu R, Hayakawa K (2005) Evaluation of estrogenic activities of hydroxylated polycyclic aromatic hydrocarbons in cigarette smoke condensate. Food Chem Toxicol 43(7):1017–1027

Keyte IJ, Harrison RM, Lammel G (2013) Chemical reactivity and long-range transport potential of polycyclic aromatic hydrocarbons – a review. Chem Soc Rev 42(24):9333–9391

Kim CS, Hu SC (1998) Regional deposition of inhaled particles in human lungs: comparison between men and women. J Appl Physiol 84(6):1834–1844

Kishikawa N, Morita S, Wada M, Ohba Y, Nakashima K, Kuroda N (2004) Determination of hydroxylated polycyclic aromatic hydrocarbons in airborne particulates by high-performance liquid chromatography with fluorescence detection. Anal Sci 20(1):129–132

Kopponen P, Törrönen R, Mäki-Paakkanen J, von Wright A, Kärenlampi S (1994) Comparison of CYP1A1 induction and genotoxicity in vitro as indicators of potentially harmful effects of environmental samples. Arch Toxicol 68(3):167–173

Kubátová A, Dronen LC, Picklo MJ, Hawthorne SB (2006) Midpolarity and nonpolar wood smoke particulate matter fractions deplete glutathione in RAW 264.7 macrophages. Chem Res Toxicol 19(2):255–261

Kubátová A, Steckler TS, Gallagher JR, Hawthorne SB, Picklo MJ Sr (2004) Toxicity of wide-range polarity fractions from wood smoke and diesel exhaust particulate obtained using hot pressurized water. Environ Toxicol Chem Int J 23(9):2243–2250

Lampi MA, Gurska J, McDonald KI, Xie F, Huang XD, Dixon DG, Greenberg BM (2006) Photoinduced toxicity of polycyclic aromatic hydrocarbons to Daphnia magna: ultraviolet-mediated effects and the toxicity of polycyclic aromatic hydrocarbon photoproducts. Environ Toxicol Chem: Int J 25(4):1079–1087

Lee HH, Choi NR, Lim HB, Yi SM, Kim YP, Lee JY (2018) Characteristics of oxygenated PAHs in PM_{10} at Seoul, Korea. Atmos Pollut Res 9(1):112–118

Lee JY, Lane DA, Heo JB, Yi SM, Kim YP (2012) Quantification and seasonal pattern of atmospheric reaction products of gas phase PAHs in $PM_{2.5}$. Atmos Environ 55:17–25

Lei Y, Shen Z, He K, Li L, Qin Y, Zeng Y, Li J, Xu H, Ma Y, Zhang T, Yang L, Zhang N, Cao J (2021) The formation and evolution of parent and oxygenated polycyclic aromatic hydrocarbons during a severe winter haze–fog event over Xi'an, China. Environ Sci Pollut Res 28(8):9165–9172

Lemieux PM, Lutes CC, Santoianni DA (2004) Emissions of organic air toxics from open burning: a comprehensive review. Prog Energy Combust Sci 30(1):1–32

Leotz-Gartziandia E, Tatry V, Carlier P (2000) Sampling and analysis of organic compounds in diesel particulate matter. Environ Monit Assess 65(1):155–163

Li N, Hao M, Phalen RF, Hinds WC, Nel AE (2003) Particulate air pollutants and asthma: a paradigm for the role of oxidative stress in PM-induced adverse health effects. Clin Immunol 109(3):250–265

Lima ALC, Farrington JW, Reddy CM (2005) Combustion-derived polycyclic aromatic hydrocarbons in the environment—a review. Environ Forensics 6(2):109–131

Liu Y, Sklorz M, Schnelle-Kreis J, Orasche J, Ferge T, Kettrup A, Zimmermann R (2006) Oxidant denuder sampling for analysis of polycyclic aromatic hydrocarbons and their oxygenated derivates in ambient aerosol: evaluation of sampling artefact. Chemosphere 62(11):1889–1898

Lundstedt, S., White, P.A., Lemieux, C.L., Lynes, K.D., Lambert, I.B., Öberg, L., ... and Tysklind, M. (2007). Sources, fate, and toxic hazards of oxygenated polycyclic aromatic hydrocarbons (PAHs) at PAH-contaminated sites. *AMBIO: A Journal of the Human Environment*, 36(6): 475-485.

Metzger, K.B., Tolbert, P.E., Klein, M., Peel, J.L., Flanders, W.D., Todd, K., ... and Frumkin, H. (2004). Ambient air pollution and cardiovascular emergency department visits. *Epidemiology*, 46-56.

Morel MC, Alers I, Arce R (2006) Photochemical degradation of 1,6- and 1,8-dinitropyrene in solution. Polycycl Aromat Compd 26(3):207–219

Nežiková B, Degrendele C, Bandowe BA, Šmejkalová AH, Kukučka P, Martiník J, Mayer L, Prokeš R, Přibylová P, Klánová J, Lammel G (2021) Three years of atmospheric concentrations of nitrated and oxygenated polycyclic aromatic hydrocarbons and oxygen heterocycles at a central European background site. Chemosphere 269:128738

Nisbet IC, Lagoy PK (1992) Toxic equivalency factors (TEFs) for polycyclic aromatic hydrocarbons (PAHs). Regul Toxicol Pharmacol 16(3):290–300

Nocun MS, Schantz MM (2013) Determination of selected oxygenated polycyclic aromatic hydrocarbons (oxy-PAHs) in diesel and air particulate matter standard reference materials (SRMs). Anal Bioanal Chem 405(16):5583–5593

Oh SM, Kim HR, Park YJ, Lee SY, Chung KH (2011) Organic extracts of urban air pollution particulate matter ($PM_{2.5}$)-induced genotoxicity and oxidative stress in human lung bronchial epithelial cells (BEAS-2B cells). Mutat Res/gen Toxicol Environ Mutagen 723(2):142–151

Ohyama M, Otake T, Adachi S, Kobayashi T, Morinaga K (2007) A comparison of the production of reactive oxygen species by suspended particulate matter and diesel exhaust particles with macrophages. Inhal Toxicol 19(sup1):157–160

Phousongphouang PT, Arey J (2003) Sources of the atmospheric contaminants, 2-nitrobenzanthrone and 3-nitrobenzanthrone. Atmos Environ 37(23):3189–3199

Schauer C, Niessner R, Pöschl U (2004) Analysis of nitrated polycyclic aromatic hydrocarbons by liquid chromatography with fluorescence and mass spectrometry detection: air particulate matter, soot, and reaction product studies. Anal Bioanal Chem 378(3):725–736

Sharma A, Singh K, Almasan A (2012) Histone H2AX phosphorylation: a marker for DNA damage. In: DNA repair protocols. Humana Press, Totowa, NJ, pp 613–626

Shen G, Chen Y, Du W, Lin N, Wang X, Cheng H, Liu J, Xue C, Liu G, Zeng EY, Xing B, Tao S (2016) Exposure and size distribution of nitrated and oxygenated polycyclic aromatic hydrocarbons among the population using different household fuels. Environ Pollut 216:935–942

Sidhu S, Gullett B, Striebich R, Klosterman J, Contreras J, DeVito M (2005) Endocrine disrupting chemical emissions from combustion sources: diesel particulate emissions and domestic waste open burn emissions. Atmos Environ 39(5):801–811

Sun J, Shen Z, Zeng Y, Niu X, Wang J, Cao J, Gong X, Xu H, Wang T, Liu H, Yang L (2018) Characterization and cytotoxicity of PAHs in $PM_{2.5}$ emitted from residential solid fuel burning in the Guanzhong Plain, China. Environ Pollut 241:359–368

Taguchi K, Fujii S, Yamano S, Cho AK, Kamisuki S, Nakai Y, Sugawara F, Froines JR, Kumagai Y (2007) An approach to evaluate two-electron reduction of 9, 10-phenanthraquinone and redox activity of the hydroquinone associated with oxidative stress. Free Radical Biol Med 43(5):789–799

Taguchi K, Shimada M, Fujii S, Sumi D, Pan X, Yamano S, Nishiyama T, Hiratsuka A, Yamamoto M, Cho AK, Froines JR, Kumagai Y (2008) Redox cycling of 9, 10-phenanthraquinone to cause oxidative stress is terminated through its monoglucuronide conjugation in human pulmonary epithelial A549 cells. Free Radical Biol Med 44(8):1645–1655

Tomaz S, Shahpoury P, Jaffrezo JL, Lammel G, Perraudin E, Villenave E, Albinet A (2016) One-year study of polycyclic aromatic compounds at an urban site in Grenoble (France): seasonal variations, gas/particle partitioning and cancer risk estimation. Sci Total Environ 565:1071–1083

Topinka J, Schwarz LR, Kiefer F, Wiebel FJ, Gajdoš O, Vidova P, Dobiáš L, Fried M, Šrám RJ, Wolff T (1998) DNA adduct formation in mammalian cell cultures by polycyclic aromatic hydrocarbons (PAH) and nitro-PAH in coke oven emission extract. Mutat Res/genet Toxicol Environ Mutagen 419(1–3):91–105

Umbuzeiro GA, Franco A, Martins MH, Kummrow F, Carvalho L, Schmeiser HH, Leykauf J, Stiborova M, Claxton LD (2008) Mutagenicity and DNA adduct formation of PAH, nitro-PAH, and oxy-PAH fractions of atmospheric particulate matter from Sao Paulo, Brazil. Mutat Res/genet Toxicol Environ Mutagen 652(1):72–80

Vione D, Barra S, De Gennaro G, De Rienzo M, Gilardoni S, Perrone MG, Pozzoli L (2004) Polycyclic aromatic hydrocarbons in the atmosphere: monitoring, sources, sinks and fate. II: Sinks and fate. Ann Chim: J Anal Environ Cult Herit Chem 94(4):257–268

Vione D, Maurino V, Minero C, Pelizzetti E, Harrison MA, Olariu RI, Arsene C (2006) Photochemical reactions in the tropospheric aqueous phase and on particulate matter. Chem Soc Rev 35(5):441–453

Vu-Duc T, Huynh CK, Binet S (2007) Laboratory generated bitumen fumes under standardized conditions. Clean-up scheme and ion trap GC-MS analysis of VOC, semi-volatile and particulate PAH and PASH. J Occup Environ Hyg 4(S1):245–248

Walgraeve C, Demeestere K, Dewulf J, Zimmermann R, Van Langenhove H (2010) Oxygenated polycyclic aromatic hydrocarbons in atmospheric particulate matter: Molecular characterization and occurrence. Atmos Environ 44(15):1831–1846

Wang J, Ho SS, Huang R, Gao M, Liu S, Zhao S, Cao J, Wang G, Shen Z, Han Y (2016) Characterization of parent and oxygenated-polycyclic aromatic hydrocarbons (PAHs) in Xi'an, China during heating period: an investigation of spatial distribution and transformation. Chemosphere 159:367–377

Webb PJ, Hamilton JF, Lewis AC, Wirtz K (2006) Formation of oxygenated-polycyclic aromatic compounds in aerosol from the photo-oxidation of o-tolualdehyde. Polycyclic Aromat Compd 26(4):237–252

Wei C, Han Y, Bandowe BAM, Cao J, Huang RJ, Ni H, Tian J, Wilcke W (2015) Occurrence, gas/particle partitioning and carcinogenic risk of polycyclic aromatic hydrocarbons and their oxygen and nitrogen containing derivatives in Xi'an, central China. Sci Total Environ 505:814–822

Wei S, Huang B, Liu M, Bi X, Ren Z, Sheng G, Fu J (2012) Characterization of $PM_{2.5}$-bound nitrated and oxygenated PAHs in two industrial sites of South China. Atmos Res 109:76–83

Wislocki PG, Wood AW, Chang RL, Levin W, Yagi H, Hernandez O, Dansette PM, Jerina DM, Conney AH (1976) Mutagenicity and cytotoxicity of benzo (a) pyrene arene oxides, phenols, quinones, and dihydrodiols in bacterial and mammalian cells. Can Res 36(9_Part_1):3350–3357

Xie M, Hannigan MP, Barsanti KC (2014) Gas/particle partitioning of *n*-alkanes, PAHs and oxygenated PAHs in urban Denver. Atmos Environ 95:355–362

Xu H, Guinot B, Cao J, Li Y, Niu X, Ho KF, Shen Z, Liu S, Zhang T, Lei Y, Zhang Q, Sun J, Gao J (2018) Source, health risk and composition impact of outdoor very fine particles (VFPs) to school indoor environment in Xi'an, Northwestern China. Sci Total Environ 612:238–246

Yu H (2002) Environmental carcinogenic polycyclic aromatic hydrocarbons: photochemistry and phototoxicity. J Environ Sci Health C 20(2):149–183

Zeng Y, Shen Z, Lei Y, Zhang T, Zhang Q, Xu H, Wang Q, Cao J, Liu Y (2018) PAHs in fine particles over Xi'an, a typical northwestern city in China: sources, distribution, and controlling factors. Environ Sci Process Impacts 20(9):1262–1272

Zhang J, Yang L, Mellouki A, Chen J, Chen X, Gao Y, Jiang P, Li Y, Yu H, Wang W (2018a) Diurnal concentrations, sources, and cancer risk assessments of $PM_{2.5}$-bound PAHs, NPAHs, and OPAHs in urban, marine and mountain environments. Chemosphere 209:147–155

Zhang J, Yang L, Mellouki A, Chen J, Chen X, Gao Y, Jiang P, Li Y, Yu H, Wang W (2018b) Atmospheric PAHs, NPAHs, and OPAHs at an urban, mountainous, and marine sites in Northern China: Molecular composition, sources, and ageing. Atmos Environ 173:256–264

Zhuo S, Du W, Shen G, Li B, Liu J, Cheng H, Xing B, Tao S (2017) Estimating relative contributions of primary and secondary sources of ambient nitrated and oxygenated polycyclic aromatic hydrocarbons. Atmos Environ 159:126–134

Environmental Microplastics, Mitochondrial Health and Human Diseases

Shashi P. Singh, Keshav Thakur, and Dhyan Chandra

1 Introduction

Plastics are synthesized from organic polymers. The convenience of plastic generation and their multi-utility properties has resulted into plastics becoming an ubiquitous part of lives of both animals and humans. Plastics have long half-lives that depend on the complexity of their structure (Chamas et al., 2020; Henkel, Hüffer and Hofmann, 2022). The half-life of a plastic may range from 500 to 1200 years (Chamas et al., 2020). Microplastics are the smaller fragments of plastic molecules. Due to lengthy half-lives, plastics do not undergo degradation easily upon disposal (Wang, Huang et al., 2021). However, under extreme physical stress conditions plastics are slowly fragmented into smaller molecules (Zhang et al., 2021). These small molecules of plastics sustain in different strata of an environment for longer duration and enter our food chain. Research has shown significant interaction among plastic and bio-molecules at cellular level. Various studies in last decade have shown the association between metabolism affiliated health risks and fragmented plastic forms (Patil et al., 2022; Sana et al., 2020; Gaur et al., 2022; Cássio et al., 2022).

Cancer is one of the most common outcomes of metabolic deregulation in the cell. Mitochondria has been identified as a pivotal component of cancer progression in cell. As a primary source of ATP production, mitochondria must essentially regulate the ROS burden, generated during the ATP production (Hibino et al., 2023). Inability of mitochondrial proteostatic machinery to keep ROS burden below threshold levels results in deregulation of mitochondrial functions, involving suppression of mitochondrial death, activation of deregulated mitochondrial fission and generation of Warburg effect in mitochondria (Kumar et al., 2022; Eckl et al., 2021). These primary

S. P. Singh · K. Thakur · D. Chandra (✉)
The Department of Pharmacology and Therapeutics, Roswell Park Comprehensive Cancer Center, Elm and Carlton Streets, Buffalo, New York 14263, USA
e-mail: dhyan.chandra@roswellpark.org

P. K. Padhy et al. (eds.), *Airborne Particulate Matter*,
https://doi.org/10.1007/978-3-031-84408-9_6

features of mitochondria contribute in survival and progression of cancer cell in a tissue.

Microplastics, due to their ability to influence the cell metabolism, have been of keen interest among researchers to examine their potential to regulate the mitochondrial function either directly or indirectly. This chapter discusses about the association between microplastics and human disease through mitochondria.

2 Microplastics

Microplastics, as described above are the smaller fragments of the large plastic molecules. They are characterized by the size ranging between 1 and 5000 μm and may exist in several forms such as fragments, films, spheres or pellets (Corella-Puertas et al., 2023; Hagelskjær et al., 2023; El Hayany et al., 2022). Microplastics are segregated from the parent compound due to various forms of physical distress involving extreme temperatures or environmental conditions (Campanale et al., 2020). However, due to extended half-lives these microplastics sustain in the environment for long time, which subsequently results in the microplastic based contamination of the food chain affecting both animals and humans alike. Due to their potential hazardous effects on health, microplastics are categorized as one of the primary health hazards for the animal and humans (Sangkham et al., 2022; Sana et al., 2020).

2.1 Properties and Application of Microplastics

Plastics are defined as the repeating unit of chemical molecule "monomer". The assembly of such monomers in various forms helps engineering the polymers with range of physical and chemical properties (Joseph et al., 2023). Plastics synthesized by combining more than one monomer is termed as copolymer for e.g. nitrile rubber or polyethylene-vinyl acetate (Gruter and Lange, 2022; Tian et al., 2022). Polymers or plastics, based on their origin/occurrence have been categorized into different forms, the most common of such polymers is petroleum based organic polymers (Siracusa and Blanco, 2020). They are used as the raw material to synthesize plastics. Synthetic polymers such as polyethylene, polyesters, nylon, epoxy are the examples of the plastics generated from petroleum based organic polymers (Nicholson et al., 2021; Bolan et al., 2020). Additionally, based on the procedure applied the polymers/plastics are subcategorized as hybrid polymers (Agatemor et al., 2023) that are composed of both organic and inorganic chemical molecules. Most common examples of hybrid polymers are rubber and silicon (Kumar et al., 2021). Plastics are composed of additives for extra thermal stability and durability. Plasticizers and flame-retardants in addition to stabilizers, reinforcing fibers and coupling agents are well-known additives used to synthesize commercial grade plastics (Ellis et al., 2021; Acquavia et al., 2021). As mentioned before the arrangement of monomers is also one of the key parameters to determine the plasticity, strength, durability and rigidness of the plastics. Plastics, based on their physical properties, are sub-categorized as semi-crystalline plastics, thermoplastics and thermosets (Desidery and Lanotte, 2022). Thermoplastics and thermosets are differentiated by basic difference among them, i.e. unlike thermoset, thermoplastics can be remolded and reshaped at high temperature.

Various forms of commonly used microplastics have been listed along with their properties and applications in Table 1.

2.2 *Prevalence of Microplastics in Environment*

Plastics are the ubiquitous part of our daily life. Inflated customer demand has resulted in significantly high production of plastics in last decade. Studies have shown that food packaging industry and automobile industry are two primary consumers of plastics (Ibrahim et al., 2021; Zhao et al., 2020). However, large scale production of plastics has resulted in overburdening of environment with plastic waste. The withering process in environment such as wind abrasion, photodegradation, ultraviolet radiation leads to generation of microplastics *via* breakdown of larger plastic molecules (Fig. 1) (Andrady et al., 2022; Hebner and Maurer-Jones, 2020; Duan et al., 2021).

Several environment-based monitoring studies have reported a significant prevalence of microplastic in the terrestrial and marine ecosystems with primary focus on marine environment contamination (Wong et al., 2020; He et al., 2022). However, the physical properties of microplastics, specifically size, make them a potential aerosol. Any anthropogenic molecule or chemical compound between the size of 2.5 and 10 μm falls under the category of aerosol (Illuminati et al., 2020). These molecules may drift with air both horizontally or vertically and can get deposited into the lungs after being inhaled by the terrestrial organisms (Sharma et al., 2023; O'Brien et al., 2023). Study from UK in 2022 using infrared spectroscopy reported the significant presence of microplastic load in the lung tissue of the lung cancer patients. The group identified 12 types of microplastics with polypropylene, polyethylene and resin contributing to the major portion of microplastic burden among cancer patients (Jenner et al., 2022). Additionally, the air borne movement of such molecules or microplastics further facilitate their distribution over the terrestrial domains of earth, especially soil. The airborne microplastic-load has been segregated into the categories of particles in indoor/outdoor air (dust) and particles in ambient air (O'Brien et al., 2023). Certain studies have shown that the microplastics load among different strata of environment is linked with each other. Marine environment has been known to be one of the most prevalent environmental reservoirs for the microplastics (Nguyen et al., 2023). Urban development near rivers and water bodies is one of the primary sources of microplastics deposition in the water bodies. Microplastic load through river gets deposited into the oceans in large concentrations. In 2015, presence of approximately 93,000–2,36,000 tonnes of plastic were reported to be found on the earth's ocean

Table 1: Microplastics type and their applications

Polymer	*Description*	*Product applications*	*References*
Polyethylene terephthalate	Polyester is a polymer derived from functional group of an ester. High purity dimethyl terephthalate and monoethylene glycol (MEG) are used to synthesize the PET. PET is categorized as thermoplastic which can be modified into different shapes by melting the material.	Polyesters are used as microfibers in food /water containers, plastic bottles, microwavable food containers. They are used as an additive for various applications in textile industry, as a resin in combination with glass fibers and are also used as synthetic fibers.	(Allen and James 2021; Ahmad, Nisar and Ahmad, 2021)
Polyamide	Polyamide is another member of thermoplastic polymer family, semi- crystalline plastic, also known as nylon. Polyamides are produced both naturally and synthetically. They are generated by repeated amide linkage.	Polyamides have high durability and strength. They are extensively used in textile industry, automative and electrical industry, for wire and cable jacketing, tire reinforcements, radiator components, fuel lines, air filters engine cover, wipers and seatbelts. Most commonly used polyamides are PA6 and PA11.	(Bahrami, Abenojar and Martínez, 2021)
Polypropylene	Polypropylene is another form of semi-crystalline plastic. It is prepared catalytically from propylene and demonstrate a significant chemical resistance and high melting point.	Due to high chemical resistance and high melting point polypropylene is extensively used in microwavable food containers, in office furniture as a flame retardant. It is used in making the clinical instrument and equipment which require chemical sterilization. It is also used as one of the major industrial derivatives in petrochemical industry.	(Xanthopoulou et al., 2021)

(continued)

Table 1: (continued)

Polymer	*Description*	*Product applications*	*References*
Polyethylene	Polethylene is categorized by two subcategories i.e. low density polyethylene and high density polyethylene. The two subcategories are divided due to their property of tensile strength and flexibility. LDPE demonstrates more flexibility, whereas HDPE has high tensile strength.	Polethylene is mostly used in packaging material, food containers, personal protection equipment kit, water and sewer pipes, garbage containers, greenhouse, electrical wiring and cables and plastic toys.	(Li et al., 2019)
Polyurethane	Polyurethane are composed of repeating units of carbamate (urethane). The applications of polyurethanes may vary depending on the different combination of monomer units. It can be used both as plasticizer and thermoplastic polymer.	Polyurethane is used as insulation foams in most residential, commercial and industrial buildings. It is used in food packaging, automobile industry and as coating material for domestic and office furniture.	(Xu et al., 2022)

(continued)

Table 1: (continued)

Polymer	*Description*	*Product applications*	*References*
Polystyrene	Polystyrene is an easily synthesized polymer. It exhibits thermoplastic properties and can withstand upto 100°C or more temperatures. It has wide applications in the medical and industrial setup.	Polystyrene is used in food packaging, automobile industry, disposable medical equipment, disposable cutlery. It is also used as an additive to other plastic polymers to enhance their thermostable properties and strength.	(Coutouly et al., 2022)
Polycarbonate	Polycarbonates are an important form of polymer, categorized into two types i.e. aromatic polycarbonates and aliphatic polycarbonates. Aromatic polycarbonates have wide industrial applications due to their thermostable properties. However, aliphatic polycarbonates are not so popular due to their ability to undergo hydrolysis easily.	Aromatic polycarbonates are called as engineering plastics. The are widely used in automobile industry, engine parts and as microwavable material in food containers. Due to high thermostability these plastics are used in industrial set-up as a mold to set up and support the equipment/ machines that function under high temperatures. Aliphatic polycarbonates are utilized extensively in medical industry due to their rapid hydrolysis property. The biodegradable property of aliphatic polycarbonate is utilized in making biodegradable implants for patients which have to undergo surgery for any form of vascular disease.	(Wei, Bhat and Darensbourg, 2023)

surface (Rakesh et al., 2020). It has been observed that in addition to human activity, certain natural phenomenon including weather, tidal waves might be significantly contributing to transferring of microplastic load from marine environment to atmospheric air. These natural phenomenon causes the plastics to disintegrate into smaller size particles ranging between the scale of μm to nm. Subsequently, the microplastics/ nano plastics gets entrapped in the water droplets, which, during extreme weather

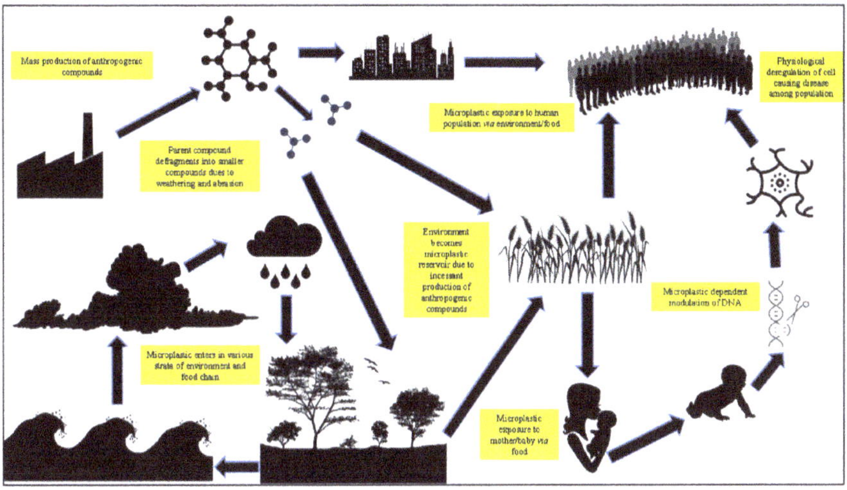

Figure 1: The prevalence of microplastics in environment and their association with physiological disorders in human population.

conditions or tidal conditions lead to bursting of microplastics containing water droplets to burst in the air. Additionally, convective updrafts cause the microplastics released into air to further move into the atmosphere. Table 2 summarizes studies from different regions of the world reporting the microplastic contamination of their environments.

2.3 Association of Microplastics with Animal/Human Non-infectious Metabolic Disorders

Accumulation of microplastics in different ecosystems of environment result in their ingestion by the inhabitants of that ecosystem, which impact the overall metabolic health of all the organisms alike. Further consumption of microplastic contaminated plant or animal-based diet by humans subsequently results in onset of several physiological diseases including cancer. In addition to food, microplastic exposure in animals and humans *via* inhalation of airborne microplastic is less popular yet an important route. Researchers in 2021 reported the presence of 13.99 µg of microplastics with the highest prevalence of high-density polyethylene in the fisherman community living on the coastal areas of Indonesia (Luqman et al., 2021). Another study from China reported the microplastic presence in placenta, meconium and infant feces. Out of 16 different types of microplastic traces, polyamide and polyurethane were significantly higher (Liu et al., 2023). Additionally, an animal study in 2019 reported the association of microplastics exposure with metabolic deregulation up to third generation of mice. In the study, female mice were exposed

Table 2: Microplastics prevalence in different geographical locations around the globe

S.No.	*Microplastic*	*Environmental strata*	*Country*	*References*
1.	Polyethylene	Sea	Greece	(Digka et al., 2018)
2.	Polyethylene, Polypropylene and polystyrene	Coastal Marine environment	Singapore	(Ng and Obbard, 2006)
3.	Polyethylene (PE), Poly(ethylene Terephthalate) (PET), and Polypropylene (PP)	Water treatment works	England	(Johnson et al., 2020)
4.	Polypropylene, Polystyrene	River	Pakistan	(Irfan et al., 2020)
5.	Polyurethane	River	Europe	(Mani et al., 2019)
6.	Polypropylene	River	China	(Zhou et al., 2020)
7.	Polyethylene (PE) and Polyethylene Terephthalate	River	India	(Amrutha and Warrier, 2020)
8.	Polyamide (PA) and Polystyrene	Coastal areas	India	(Ranjani et al., 2021)
9.	Polyethylene, Polypropylene, Polyvinyl chloride, Polyethylene terephthalate, and Polyvinyl acetate	Soil	India	(Singh et al., 2023)
10.	Polyethylene, Polypropylene, Polyamides, Polyethylene terephthalate	Coastal area	Italy	(Scopetani et al., 2021)
11.	Polypropylene (PP) Polyethylene (PE) and Polyethylene terephthalate	Water bodies	Turkey	(Karaoğlu and Gül, 2020)
12.	Polyethylene, Polypropylene, Polystyrene, Polyamide, and Polyvinyl chloride	Soil	China	(Zhou, Liu and Wang, 2019)
13.	Polyethylene and Polypropylene	Harbour	Jamaica	(Rose and Webber, 2019)
14.	Polyvinyl chloride, Polypropylene, Polyethylene, Polyester, and Polyamide	Water body	Italy	(Pellini et al., 2018)

to polystyrene microplastic through food during gestation and lactation periods. Constant xenocompound exposure resulted in the gut microbiota dysbiosis and gut barrier dysfunction in the first and second progeny. The study demonstrated the potential long-term risk associated with microplastic exposure (Luo et al., 2019). Another animal study showed that microplastics exposure through food also results in insulin resistance and developing diabetes in the test animals (Shi et al., 2022). Also, the animal studies have shown the association of microplastics exposure in animals with non-alcoholic fatty liver disease (NAFLD), obesogens as well as dereg-ulation in lipid and energy metabolism (Marra and Svegliati-Baroni, 2018; Mouries et al., 2019). Microplastics, in addition to metabolic health also influence the genetic stability of both animals and humans. Microplastics have been found to have affinity towards guanine. Guanine attached to ribose ring, when phosphorylated results in the byproducts that are essential to maintain the biochemical process of nucleic acid and protein stability in cells. However, microplastics (due to their affinity towards guanine) are understood to generate mutation hotspots near or in the guanine rich regions in gene. Such affinity of microplastics towards guanine in-turn, may result in hampering guanine's role of nucleic acid and protein stability in the cells. These mutational hotspots have been identified in the patient cohorts signifying the role of microplastics in the onset and regulation of cancer in both animals and humans (Hu et al., 2021). Animal study showed that exposure to microplastics also induce chemo resistance among animal models that developed cancer upon dietary exposure of microplastics (Kim et al., 2022; Nabi et al., 2022). A study in zebra fish model showed the influence of microplastics over the mitochondria regulation and size especially in brain and liver. Chronic exposure of microplastics to zebra fish resulted in deterioration of complex II and IV, deficiency of mitochondrial respiration and loss of mitochondrial membrane potential. The study recorded the increased reactive oxygen species (ROS) load among the liver and brain cells leading to mitochondrial dysfunction (Félix et al., 2023). Association of microplastics with mitochondrial dysfunction and ROS generation has also been shown in *in vitro* studies. Study reported the significant modulation of redox related genes in the mitochondrial niche when cells were exposed to spherical or fiber shaped polystyrene for 24 hours (Saenen et al., 2023). In addition to brain and liver, microplastic-based deregulation of mitochondrial homeostasis and endoplasmic reticulum stress has also been reported in the kidney cells, spermatocyte and neuronal cells in both humans and mice models (Wang, Lee, et al., 2021; Liu, Hou, et al., 2022; Wu et al., 2023; Zhang, Sun, et al., 2022).

3 Mitochondria

Mitochondria play a significant role in energy metabolism. It supplies around 90% of ATP through oxidative phosphorylation for cellular functioning and growth. Thus, it is also referred to as the powerhouse of cells. Mitochondria are involved in various cellular processes like regulation of redox levels, homeostasis, cell signalling and cell death mechanisms. They play a central role in cell survival and apoptosis. They are involved in many diseases, like neurologic diseases, metabolic diseases, and cancers.

Mitochondria arise from fusion of prokaryotic structures into eukaryotic cells, later forming endosymbiotic associations. Thus, it has its own DNA which transcribes around 13 major mitochondrial related proteins of oxidative phosphorylation (OxPhos). It requires constant signal exchange between the nucleus and cytosol. Any discrepancy in mtDNA leads to mitochondrial dysfunctions. Generally, in cells, the number and amount of mitochondrial DNA varies. Interestingly, the importance of mitochondrial DNA was well studied in Tasmanian devils infected with Canine transmissible venereal tumor (CTVT). From generation to generation, the tumor was transferred from one animal to another, leading to decay of mtDNA. However, the procurement of mtDNA from transient hosts makes it functional and thus promoting CTVT cell growth. Recently, studies have shown that replication of mitochondrial DNA is probably independent of nuclear control. Many microplastics are involved in the modulation of mtDNA replication.

Besides energy production, mitochondria also play an important role in reactive oxygen production. The electron transport chain (ETC) is present in the inner membrane of mitochondria where mitochondrial ROS is produced.

4 Impact of Microplastics on Mitochondria

The use of plastics has eased our life. However, excessive usage of plastics in food grade categories plays a hazardous role in affecting human health. Latest research shows that mitochondria were modulated by microplastics and impacts human health. The precise mechanism of how microplastics regulate the mitochondria is poorly understood but the probable mechanism is influencing mitochondrial dysfunction, inflammation, mtDNA damage response, and endocrine disrupting properties (Mani et al., 2023). Mitochondria also play a major role in synthesizing the lipid species and generating the metabolic substrate for post-translational modifications that play a key role in various signalling. Therefore, the mitochondria and nucleus must communicate to effectively coordinate the mitochondrial homeostasis and respond to the demands (Wallace, 2012).

4.1 Microplastics Induced Mitochondrial Dysfunction

Mitochondrial dysfunctions are associated with lack of ATP generation, metabolic production, import/export of proteins, which leads to misbalancing the electron transport chain. Mitochondria are not static organelles. They exist in dynamic environments and different architectures. Therefore, mitochondrial number (quantity) varies from cell to cell. Although any abrupt change in numbers also leads to mito dysfunction. Mito dysfunction results in fragmentation, fusion and mitophagy (Vyas et al., 2016; Wallace, 2012).

Mito fusion is a typical fusion reaction where two mitochondria collide and mix their components to form a single mitochondrion. Mito fusion depends on GTPases like MFN1-2 and OPA1. The fusion is a two-step process where the first step is fusion of the outer mitochondrial membrane (OMM) followed by inner mitochondrial membrane (IMM) fusion. Mitofusing (Mfn-1/2) protein plays a key role in fusing the outer mito membrane of mitochondria. Studies have shown that overexpression of the Mfn-1/2 leads to clustering of mitochondria. MFN molecules interact with the OMM of other mitochondria, ultimately forming the homo- or hetero-mitochondrial complexes by utilizing the GTP (Vyas et al., 2016).

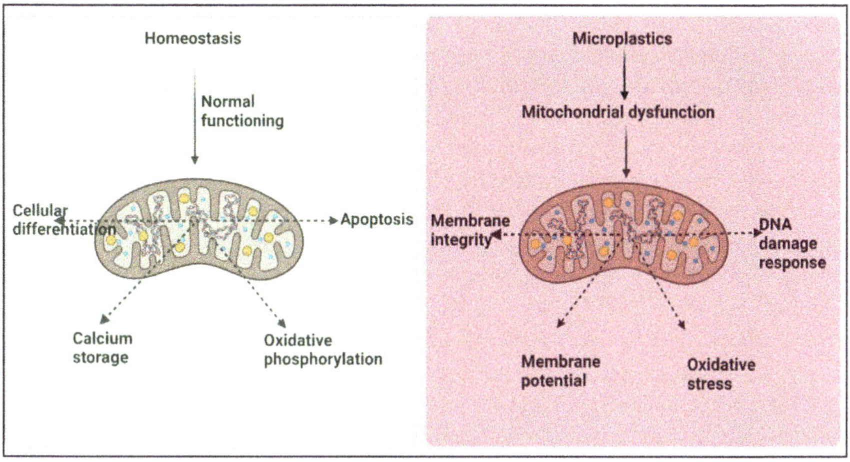

Figure 2: Role of mitochondria and microplastic induced dysfunction in mitochondria.

Mitochondrial OMM fusion is independent of the inner mitochondrial membrane. Optic atrophy 1 (OPA1) is a GTPase which plays a critical role in fusing the IMM. The OPA1 interacts with Cardiolipin (CL) to finally stitch the IMM together. Mitochondrial fission is executed by Dynamin related proteins (Drps). The DRP1 and DNM2 play major roles in fragmenting the mitochondria. Initially, DRP1 localizes in cytosol by post-translational regulation, but after stimulation, it travels to the mitochondrial outer surface of mitochondria. Studies have shown that DRP1 is regulated by ubiquitination, phosphorylation-dephosphorylation, nitrosylation and glycosylation in a cell-specific fashion. Interestingly the phosphorylation and dephosphorylation drive the DRP1 in a bidirectional way. The phosphorylation at the Ser637 site on DRP1 leads to accumulation in cytosol while dephosphorylation by calcineurin leads to mitochondrial membrane localization. After the activation, DRP1 oligomerizes, forming construction of the mito tuber. The fission process is dependent on many factors which play important roles. Any disruption in key signalling molecules or pathological stress leads to failure of fission, finally leading to mitochondrial dysfunction (Fig. 2) (Zong et al., 2016). The dysfunction is attributed to many physiological disorders like fatigue, neurodegenerative disorders and metabolic disorders.

4.2 *Microplastics Modulate the Mitochondrial ROS*

Mitochondria are the primary site of oxidative phosphorylation and the major location for ROS production. ROS plays a crucial role in health and maintaining homeostasis. However, high concentration damages the organelles, membranes, mito

DNA, genomic DNA and proteins, leading to mitochondrial dysfunction. Dysfunctional mitochondria either produce high levels of ROS or do not produce them irrespective of stimulation. The generation of ROS is the by-product of oxidative phosphorylation. Another mechanism of ROS creation is the induction of uncoupling proteins which are finally involved in leakage of proton gradients from the inner mitochondrial membrane into the matrix. The leakage leads to reduction in ATP formation while consuming oxygen (Nicolson, 2014). Studies have shown that ageing has been associated with deterioration of antioxidant defense efficiency, which leads to the accumulation of ROS, leading to continued oxidative stress. The long exposure of microplastic components can lead to chronic inflammation and finally lead to alteration in subcellular ROS like mito ROS, ER ROS. In *Drosophila melanogaster*, it was observed that overexpressing the antioxidant enzymes leads to efficient removal of ROS, thus protecting the DNA and integrity of cell (Towarnicki et al., 2020). The ROS includes hydroxyl radicals (*OH) which are more reactive than superoxide (O_2*-) and hydrogen peroxide (H_2O_2). Likewise, reactive nitrogen species (RNS) mediates nitric oxide (NO*) are less reactive than peroxynitrite ($ONOO^-$) (Brookes and Darley-Usmar, 2004; Radi et al., 2002). The generation of reactive oxygen species is a continuous process and requires for maintaining the homeostasis (Kowaltowski et al., 2009; Murphy, 2009).

Mitochondria plays a highly efficient antioxidant defense system which protects itself from ROS and thus plays a significant role in scavenging the ROS from the system. However, when the overwhelming amount of ROS is produced and the mito cannot scavenge, it damages the mito DNA more, as compared to nuclear DNA, as mito have less efficient repair methods. It is more susceptible to ROS as it is located on the inner membrane of the mito and thus present in the immediate vicinity of ROS. Mito DNA is associated with many proteins which are required for maintaining the mito DNA. The ROS disturbs this interaction and leads to mito DNA damage and finally mito dysfunctions (Venditti and Di Meo, 2020).

4.3 Microplastics Influencing the Mitochondrial Dynamics

Micro- and nano-plastics (MNPs) are spread everywhere and play a crucial role in day-to-day life. They are artificial polymers with sizes $\leq$ 5 mm (Microplastics-MPs) and 1 nm to 1 μm for nanoplastics (NPs). They are non-biodegradable and thus do not produce end products. Microplastics enter the environment via food packaging. Studies have shown that MP was reported in chicken stomachs and feces. MNPs are internalized by cells and lead to rupture or injuring the cells. MNPs were found to modulate the cell viability, ROS level, mito membrane potential. Lin et al. (2022) showed that MNPs induce mitochondrial damage, mitochondrial ROS and inhibition of mitochondrial respiration. Another study by Shen et al. (2022) showed that accumulation of microplastics induces hepatic damage by activating the cGAS/STING pathway, finally leading to upregulating the pro-inflammatory cytokines. Yin et al. (2023) found that microplastics induce inflammation in the liver and stimulate extracellular traps in macrophages, causing liver injury. They found that microplastics were actively phagocytosed by macrophages, which disrupts mitochondrial homeostasis. In the HepG2 cell line, nanoplastics showed induction of caspase-3, caspase-9, Cytochrome C (Li et al., 2023). Table 3 summarize *in vitro* and *in vivo* studies conducted to assess the impact of microplastics over mitochondrial function.

Microplastics can not only cause damage to tissues, but their role in spermatogenesis is also found. Liu et al. found that in mouse spermatocytes line, GC-2 cells, microplastics induce mitochondrial damage by decreasing ATP content, mito genome integrity and reduced mito membrane potential and leading to mitochondrial PINK1/ Parkin autophagy activation (Liu, Hou et al., 2022). The role of microplastics was also studied in C elegans. Studies by Qu et al. (2023) showed that nanoplastics can inhibit mitochondrial unfolded protein responses (UPR) by regulating the hsp6, ubl-5, dve-1 and clpp-1 and promoting mitochondrial apoptosis. Latest reports showed that microplastics can activate a plethora of signaling cascades that can eventually lead to DNA damage, protein oxidation, lipid peroxidation and promote pulmonary toxicity, hepatic injury, cardiac dysfunction, neurotoxicity, immunotoxicity and reproductive dysfunctions. Although a large magnitude of research will be required to assess the role of microplastics in humans and the necessity of a suitable model system will be needed to investigate its response in humans and other life forms.

5 Conclusions

As human population is rising, so is ubiquitous presence of plastics in daily life due to its ever-increasing demand and economical affordability. However, the improper handling and abuse of plastics has resulted in creating non-biodegradable pollution resulting in microplastics contaminating water, soil and air ecosystems. Number of environmental monitoring studies have reported the presence of microplastics load among population of variable geographical regions. These studies also suggest that micro and nano plastics once disposed in environment do not remain in same location, instead they may reach to other places through air or water bodies. Several studies have reported microplastic associated toxicity due to their high bioaccumulation potential in both animals and humans. Microplastics are also linked with various metabolic health problems in every life form. Most studies show that microplastics impacts on mitochondrial functions. They enhance oxidative stress and mitochondrial membrane potential, leading to mitochondrial dysfunction. However, their role in mitochondrial dysfunction needs to be explored intensively and should not be ignored.

Table 3: Microplastics impact on mitochondrial functions

Model	*Plastic category*	*Mechanism*	*References*
In vitro model			
Mouse GC-2	Polystyrene microplastics	Oxidative stress in mitochondrial structure	(Liu, Hou et al., 2022)
Human gastric epithelial (GES-1) cells	Polystyrene microplastics	PI3K/AKT mediated oxidative stress in mitochondria	(Qin et al., 2022)
HL7702 hepatocytes	Microplastics	PPARα and PGC-1α expression mitochondrial damage and dysfunction in liver	(Shen et al., 2022)
Human L02 hepatocytes and C57BL/6 mice	Polystyrene microplastics	PERK gene influences mitochondrial fission/ fusion	(Woo et al., 2023)
Murine macrophages	Microplastics	Impact on glycolysis and reduction in mitochondrial respiration.	(Merkley et al., 2022)
Human colonic epithelial cell CCD841CoN and small intestine epithelial cell HIEC-6	Polystyrene microspheres (PS-MPs) and nanospheres (PS-NPs)	Modulating the mitochondrial mebrane potential and oxidative stress	(Zhang, Wang et al., 2022)
In vivo model			
Daphnia carinata	Microplastics	Aberration in apoptosis and ferroptosis pathways	(Jiang et al., 2023)
Caenorhabditis elegans	Nanoplastics	Dysregulation of mitochondrial homeostasis	(Qu et al., 2023)
Specific pathogen free (SPF) CD-1® mice	Polystyrene microplastics	Reduced mitochondrial membrane potential and increased ROS	(Liu, Zhuan et al., 2022)
Zebra fish	Polyethylene microplastics	Increased intestinal inflammation	(Tarasco et al., 2022)
Zebra fish	Microplastics	Decrease in mitochondrial respiratory chain complex II and IV activity	(Félix et al., 2023)
Chickens	Polystyrene microplastics	Cardiotoxicity caused by mitochondrial dysfunction	(Zhang, Yin et al., 2022)

References

Acquavia MA, Pascale R, Martelli G, Bondoni M, Bianco G (2021) Natural polymeric materials: a solution to plastic pollution from the agro-food sector. Polymers 13:158

Agatemor C, Ghosal K, Fura S, Foot PJS (2023) Fundamentals of inorganic and organometallic polymer science, 1st edn. CPI Group (UK) Ltd., Croydon, United Kingdom

Ahmad F, Nisar H, Ahmad N (2021) Simulation and optimization of polymerization reactor for the production of polyethylene terephthalate (PET). Austin Chem Eng 8:1083

Allen RD, James MI (2021) Chemical recycling of PET. Circular economy of polymers: topics in recycling technologies. ACS Publications

Amrutha K, Warrier AK (2020) The first report on the source-to-sink characterization of microplastic pollution from a riverine environment in tropical India. Sci Total Environ 739:140377

Andrady AL, Barnes PW, Bornman JF, Gouin T, Madronich S, White CC, Zepp RG, Jansen MAK (2022) Oxidation and fragmentation of plastics in a changing environment: from UV-radiation to biological degradation. Sci Total Environ 851:158022

Bahrami M, Abenojar J, Martínez MA (2021) Comparative characterization of hot-pressed polyamide 11 and 12: Mechanical, thermal and durability properties. Polymers 13:3553

Bolan NS, Karthikeyan K, Bolan SS, Kirkham MB, Kim K-H, Hou D (2020) An introduction to the chemistry and manufacture of plastics, vol 1, pp 85–93

Brookes PS, Darley-Usmar VM (2004) Role of calcium and superoxide dismutase in sensitizing mitochondria to peroxynitrite-induced permeability transition. Am J Physiol - Heart Circ Physiol 286:H39–H46

Campanale C, Massarelli C, Savino I, Locaputo V, Uricchio VF (2020) A detailed review study on potential effects of microplastics and additives of concern on human health. Int J Environ Res Public Health 17:1212

Cássio F, Batista D, Pradhan A (2022) Plastic interactions with pollutants and consequences to aquatic ecosystems: what we know and what we do not know. Biomolecules 12:798

Chamas A, Moon H, Zheng J, Qiu Y, Tabassum T, Jang JH, Abu-Omar M, Scott SL, Suh S (2020) Degradation rates of plastics in the environment. ACS Sustain Chem & Eng 8:3494–3511

Corella-Puertas E, Hajjar C, Lavoie J, Boulay A-M (2023) MarILCA characterization factors for microplastic impacts in life cycle assessment: physical effects on biota from emissions to aquatic environments. J Clean Prod 418:138197

Coutouly C, Van Ruymbeke E, Ramos L, Dieudonné-George P, Fustin C-A (2022) Exploiting the lower disorder-to-order temperature in polystyrene-b-poly (n-butyl acrylate)-b-polystyrene triblock copolymers to increase their flow resistance at high temperature. J Rheol 66:1305–1318

Desidery L, Lanotte M (2022) Polymers and plastics: types, properties and manufacturing. Plastic waste for sustainable asphalt roads. Elsevier

Digka N, Tsangaris C, Kaberi H, Adamopoulou A, Zeri C (2018) Microplastic abundance and polymer types in a Mediterranean environment. In: Proceedings of the international conference on microplastic pollution in the Mediterranean Sea. Springer, pp 17–24

Jiajun D, Bolan N, Li Y, Ding S, Atugoda T, Vithanage M, Sarkar B, Tsang DCW, Kirkham MB (2021) Weathering of microplastics and interaction with other coexisting constituents in terrestrial and aquatic environments. Water Res 196:117011

Eckl E-M, Ziegemann O, Krumwiede L, Fessler E, Jae LT (2021) Sensing, signaling and surviving mitochondrial stress. Cell Mol Life Sci 78:5925–5951

El Hayany B, Rumpel C, Hafidi M, El Fels L (2022) Occurrence, analysis of microplastics in sewage sludge and their fate during composting: a literature review. J Environ Manag 317:115364

Ellis LD, Rorrer NA, Sullivan KP, Otto M, McGeehan JE, Román-Leshkov Y, Wierckx N, Beckham GT (2021) Chemical and biological catalysis for plastics recycling and upcycling. Nat Catal 4:539–556

Félix L, Carreira P, Peixoto F (2023) Effects of chronic exposure of naturally weathered microplastics on oxidative stress level, behaviour, and mitochondrial function of adult zebrafish (Danio rerio). Chemosphere 310:136895

Gaur VK, Gupta S, Sharma P, Gupta P, Varjani S, Srivastava JK, Chang J-S, Bui X-T (2022) Metabolic cascade for remediation of plastic waste: a case study on microplastic degradation. Curr Pollut Rep 8:30–50

Gruter G-JM, Lange J-P (2022). Tutorial on polymers – manufacture, properties, and applications. In: Biodegradable polymers in the circular plastics economy, pp 83–111

Hagelskjær O, Crézé A, Le Roux G, Sonke JE (2023) Investigating the correlation between morphological features of microplastics (5–500 μm) and their analytical recovery. Microplastics Nanoplastics 3:1–13

He B, Liu An, Duan H, Wijesiri B, Goonetilleke A (2022) Risk associated with microplastics in urban aquatic environments: a critical review. J Hazard Mater 439:129587

Hebner TS, Maurer-Jones MA (2020) Characterizing microplastic size and morphology of photodegraded polymers placed in simulated moving water conditions. Environ Sci Process Impacts 22:398–407

Henkel C, Hüffer T, Hofmann T (2022) Polyvinyl chloride microplastics leach phthalates into the aquatic environment over decades. Environ Sci Technol 56:14507–14516

Hibino M, Maeki M, Tokeshi M, Ishitsuka Y, Harashima H, Yamada Y (2023) A system that delivers an antioxidant to mitochondria for the treatment of drug-induced liver injury. Sci Rep 13:6961

Hu X, Biswas A, Sharma A, Sarkodie H, Tran I, Pal I, De S (2021) Mutational signatures associated with exposure to carcinogenic microplastic compounds bisphenol A and styrene oxide. NAR Cancer 3:zcab004

Ibrahim NI, Shahar FS, Sultan MTH, Ain Umaira Md, Shah SN, Safri A, Yazik MHM (2021) Overview of bioplastic introduction and its applications in product packaging. Coatings 11:1423

Illuminati S, Annibaldi A, Bau S, Scarchilli C, Ciardini V, Grigioni P, Girolametti F, Vagnoni F, Scarponi G, Truzzi C (2020) Seasonal evolution of size-segregated particulate mercury in the atmospheric aerosol over Terra Nova Bay, Antarctica. Molecules 25:3971

Irfan M, Qadir A, Mumtaz M, Ahmad SR (2020) An unintended challenge of microplastic pollution in the urban surface water system of Lahore, Pakistan. Environ Sci Pollut Res 27:16718–16730

Jenner LC, Rotchell JM, Bennett RT, Cowen M, Tentzeris V, Sadofsky LR (2022) Detection of microplastics in human lung tissue using μFTIR spectroscopy. Sci Total Environ 831:154907

Jiang Y, Li Q, Wang Y, Jin J, Wei W, Zhang Y, Yang H (2023) Polyester microplastic fibers induce mitochondrial damage, apoptosis and oxidative stress in Daphnia carinata, accompanied by changes in apoptotic and ferroptosis pathway. Aquat Toxicol 263:106690

Johnson AC, Ball H, Cross R, Horton AA, Jürgens MD, Read DS, Vollertsen J, Svendsen C (2020) Identification and quantification of microplastics in potable water and their sources within water treatment works in England and Wales. Environ Sci Technol 54:12326–12334

Joseph TM, Hasanin MS, Unni AB, Mahapatra DK, Haponiuk J, Thomas S (2023) Macromolecules: contemporary futurist thoughts on progressive journey. Eng 4:678–702

Kaan K, Gül S (2020) Characterization of microplastic pollution in tadpoles living in small waterbodies from Rize, the northeast of Turkey. Chemosphere 255:126915

Kim H, Zaheer J, Choi E-J, Kim JS (2022) Enhanced ASGR2 by microplastic exposure leads to resistance to therapy in gastric cancer. Theranostics 12:3217

Kowaltowski AJ, de Souza-Pinto NC, Castilho RF, Vercesi AE (2009) Mitochondria and reactive oxygen species. Free Radic Biol Med 47:333–343

Kumar R, Chaudhary AK, Woytash J, Inigo JR, Gokhale AA, Bshara W, Attwood K, Wang J, Spernyak JA, Rath E, Yadav N (2022) A mitochondrial unfolded protein response inhibitor suppresses prostate cancer growth in mice via HSP60. J Clin Investig 132:e149906

Kumar V, Alam MN, Manikkavel A, Choi J, Lee DJ (2021) Investigation of silicone rubber composites reinforced with carbon nanotube, nanographite, their hybrid, and applications for flexible devices. J Vinyl Add Tech 27:254–263

Li D, Zhou L, Wang X, He L, Yang X (2019) Effect of crystallinity of polyethylene with different densities on breakdown strength and conductance property. Materials 12:1746

Li Y, Guo M, Niu S, Shang M, Chang X, Sun Z, Zhang R, Shen X, Xue Y (2023) ROS and DRP1 interactions accelerate the mitochondrial injury induced by polystyrene nanoplastics in human liver HepG2 cells. Chem Biol Interact 379:110502
Lin S, Zhang H, Wang C, Xiu-Li Su, Song Y, Pengfei Wu, Yang Z, Wong M-H, Cai Z, Zheng C (2022) Metabolomics reveal nanoplastic-induced mitochondrial damage in human liver and lung cells. Environ Sci Technol 56:12483–12493
Liu S, Guo J, Liu X, Yang R, Wang H, Sun Y, Chen Bo, Dong R (2023) Detection of various microplastics in placentas, meconium, infant feces, breastmilk and infant formula: a pilot prospective study. Sci Total Environ 854:158699
Liu T, Hou B, Wang Z, Yang Y (2022) Polystyrene microplastics induce mitochondrial damage in mouse GC-2 cells. Ecotoxicol Environ Saf 237:113520
Liu Z, Zhuan Q, Zhang L, Meng L, Xiangwei Fu, Hou Y (2022) Polystyrene microplastics induced female reproductive toxicity in mice. J Hazard Mater 424:127629
Luo T, Wang C, Pan Z, Jin C, Zhengwei Fu, Jin Y (2019) Maternal polystyrene microplastic exposure during gestation and lactation altered metabolic homeostasis in the dams and their F1 and F2 offspring. Environ Sci Technol 53:10978–10992
Luqman A, Nugrahapraja H, Wahyuono RA, Islami I, Haekal MH, Fardiansyah Y, Putri BQ, Amalludin FI, Rofiqa EA, Götz F (2021) Microplastic contamination in human stools, foods, and drinking water associated with Indonesian coastal population. Environments 8:138
Mani S, Ralph SJ, Swargiary G, Rani M, Wasnik S, Singh SP, Devi A (2023) Therapeutic targeting of mitochondrial plasticity and redox control to overcome cancer chemoresistance. Antioxid Redox Signal 39:591–619
Mani T, Primpke S, Lorenz C, Gerdts G, Burkhardt-Holm P (2019) Microplastic pollution in benthic midstream sediments of the Rhine River. Environ Sci Technol 53:6053–6062
Marra F, Svegliati-Baroni G (2018) Lipotoxicity and the gut-liver axis in NASH pathogenesis. J Hepatol 68:280–295
Merkley SD, Moss HC, Goodfellow SM, Ling CL, Meyer-Hagen JL, Weaver J, Campen MJ, Castillo EF (2022) Polystyrene microplastics induce an immunometabolic active state in macrophages. Cell Biol Toxicol 38:31–41
Mouries J, Brescia P, Silvestri A, Spadoni I, Sorribas M, Wiest R, Mileti E, Galbiati M, Invernizzi P, Adorini L (2019) Microbiota-driven gut vascular barrier disruption is a prerequisite for non-alcoholic steatohepatitis development. J Hepatol 71:1216–1228
Murphy MP (2009) How mitochondria produce reactive oxygen species. Biochem J 417:1–13
Nabi G, Ahmad S, Ullah S, Zada S, Sarfraz M, Guo X, Ismail M, Wanghe K (2022) The adverse health effects of increasing microplastic pollution on aquatic mammals. J King Saud Univ - Sci 34:102006
Ng KL, Obbard JP (2006) Prevalence of microplastics in Singapores coastal marine environment. Mar Pollut Bull 52:761–767
Nguyen M-K, Lin C, Nguyen H-L, Le Van-Re KL, Priya JS, Chang SW, Um M-J, Duc Nguyen D (2023) Emergence of microplastics in the aquatic ecosystem and their potential effects on health risks: the insights into Vietnam. J Environ Manag 344:118499
Nicholson SR, Rorrer NA, Carpenter AC, Beckham GT (2021) Manufacturing energy and greenhouse gas emissions associated with plastics consumption. Joule 5:673–686
Nicolson GL (2014) Mitochondrial dysfunction and chronic disease: treatment with natural supplements. Integr Med: Clin J 13:35
O'Brien S, Rauert C, Ribeiro F, Okoffo ED, Burrows SD, O'Brien JW, Wang X, Wright SL, Thomas KV (2023) Theres something in the air: a review of sources, prevalence and behaviour of microplastics in the atmosphere. Sci Total Environ 874:162193
Patil PB, Maity S, Sarkar A (2022) Potential human health risk assessment of microplastic exposure: Current scenario and future perspectives. Environ Monit Assess 194:898
Pellini G, Gomiero A, Fortibuoni T, Ferrà C, Grati F, Tassetti AN, Polidori P, Fabi G, Scarcella G (2018) Characterization of microplastic litter in the gastrointestinal tract of Solea solea from the Adriatic Sea. Environ Pollut 234:943–952

Qin J, Xia P-F, Yuan X-Z, Wang S-G (2022) Chlorine disinfection elevates the toxicity of polystyrene microplastics to human cells by inducing mitochondria-dependent apoptosis. J Hazard Mater 425:127842

Man Q, Miao L, Chen H, Zhang X, Wang Y (2023) SKN-1/Nrf2-dependent regulation of mitochondrial homeostasis modulates transgenerational toxicity induced by nanoplastics with different surface charges in Caenorhabditis elegans. J Hazard Mater 457:131840

Radi R, Cassina A, Hodara R, Quijano C, Castro L (2002) Peroxynitrite reactions and formation in mitochondria. Free Radical Biol Med 33:1451–1464

Rakesh SS, Ramasamy M, Ramesh PT, Maheswari M, Shri Rangasami SR, Yuvaraj M (2020) Nanoplastics in 21st Century. Biot Res Today 2:356–358

Ranjani M, Veerasingam S, Venkatachalapathy R, Mugilarasan M, Bagaev A, Mukhanov V, Vethamony PJMPB (2021) Assessment of potential ecological risk of microplastics in the coastal sediments of India: a meta-analysis. Mar Pollut Bull 163:111969

Rose D, Webber M (2019) Characterization of microplastics in the surface waters of Kingston Harbour. Sci Total Environ 664:753–760

Saenen ND, Witters MS, Hantoro I, Tejeda I, Ethirajan A, Van Belleghem F, Smeets K (2023) Polystyrene microplastics of varying sizes and shapes induce distinct redox and mitochondrial stress responses in a caco-2 monolayer. Antioxidants 12:739

Sana SS, Dogiparthi LK, Gangadhar L, Chakravorty A, Abhishek N (2020) Effects of microplastics and nanoplastics on marine environment and human health. Environ Sci Pollut Res 27:44743–44756

Sangkham S, Faikhaw O, Munkong N, Sakunkoo P, Arunlertaree C, Chavali M, Mousazadeh M, Tiwari A (2022) A review on microplastics and nanoplastics in the environment: their occurrence, exposure routes, toxic studies, and potential effects on human health. Mar Pollut Bull 181:113832

Scopetani C, Chelazzi D, Martellini T, Pellinen J, Ugolini A, Sarti C, Cincinelli A (2021) Occurrence and characterization of microplastic and mesoplastic pollution in the Migliarino San Rossore, Massaciuccoli Nature Park (Italy). Mar Pollut Bull 171:112712

Sharma S, Bhardwaj A, Thakur M, Saini A (2023) Understanding microplastic pollution of marine ecosystem: a review. Environ Sci Pollut Res 31:1–44

Shen R, Yang K, Cheng X, Guo C, Xing X, Sun H, Liu D, Liu X, Wang D (2022) Accumulation of polystyrene microplastics induces liver fibrosis by activating cGAS/STING pathway. Environ Pollut 300:118986

Shi C, Han X, Guo W, Qi Wu, Yang X, Wang Y, Tang G, Wang S, Wang Z, Liu Y (2022) Disturbed gut-liver axis indicating oral exposure to polystyrene microplastic potentially increases the risk of insulin resistance. Environ Int 164:107273

Singh S, Chakma S, Alawa B, Kalyanasundaram M, Diwan V (2023) Identification, characterization, and implications of microplastics in soil – a case study of Bhopal, central India. J Hazard Mater Adv 9:100225

Valentina S, Blanco I (2020) Bio-polyethylene (Bio-PE), Bio-polypropylene (Bio-PP) and Bio-poly (ethylene terephthalate)(Bio-PET): recent developments in bio-based polymers analogous to petroleum-derived ones for packaging and engineering applications. Polymers 12:1641

Tarasco M, Gavaia PJ, Bensimon-Brito A, Cordelieres FP, Santos T, Martins G, De Castro DT, Silva N, Cabrita E, Bebianno MJ (2022) Effects of pristine or contaminated polyethylene microplastics on zebrafish development. Chemosphere 303:135198

Tian W, Song P, Zhang H, Duan X, Wei Y, Wang H, Wang S (2022) Microplastic materials in the environment: problem and strategical solutions. Prog Mater Sci 132:101035

Towarnicki SG, Kok LM, William J, Ballard O (2020) Yin and Yang of mitochondrial ROS in Drosophila. J Insect Physiol 122:104022

Paola V, Di Meo S (2020) The role of reactive oxygen species in the life cycle of the mitochondrion. Int J Mol Sci 21:2173

Vyas S, Zaganjor E, Haigis MC (2016) Mitochondria and cancer. Cell 166:555–566

Wallace Douglas C (2012) Mitochondria and cancer. Nat Rev Cancer 12:685–698

Wang G-X, Huang D, Ji J-H, Völker C, Wurm FR (2021) Seawater-degradable polymers—fighting the marine plastic pollution. Adv Sci 8:2001121

Wang Y-L, Lee Y-H, Hsu Y-H, Chiu I-J, Huang CC-Y, Huang C-C, Chia Z-C, Lee C-P, Lin Y-F, Chiu H-W (2021) The kidney-related effects of polystyrene microplastics on human kidney proximal tubular epithelial cells HK-2 and male C57BL/6 mice. Environ Health Perspect 129:057003

Wei P, Bhat GA, Darensbourg DJ (2023) Enabling new approaches: recent advances in processing aliphatic polycarbonate-based material. Angew Chem Int Ed 62:e202307507

Wong JKH, Lee KK, Tang KHD, Yap P-S (2020) Microplastics in the freshwater and terrestrial environments: prevalence, fates, impacts and sustainable solutions. Sci Total Environ 719:137512

Woo J-H, Seo HJ, Lee J-Y, Lee I, Jeon K, Kim B, Lee K (2023) Polypropylene nanoplastic exposure leads to lung inflammation through p38-mediated NF-κB pathway due to mitochondrial damage. Part Fibre Toxicol 20:1–17

Wu D, Zhang M, Bao TT, Lan H (2023) Long-term exposure to polystyrene microplastics triggers premature testicular aging. Part Fibre Toxicol 20:1–16

Xanthopoulou E, Terzopoulou Z, Zamboulis A, Papadopoulos L, Tsongas K, Tzetzis D, Papageorgiou GZ, Bikiaris DN (2021) Poly (propylene vanillate): a sustainable lignin-based semicrystalline engineering polyester. ACS Sustain Chem & Eng 9:1383–1397

Xu J, Xiao W, Zhang S, Dong Z, Lei C (2022) Synthesis and characterization of polyurethane with poly (ether-ester) diols soft segments consisted by ether and ester linkages in one repeating unit. Eur Polym J 179:111553

Yin K, Wang D, Zhang Y, Hongmin Lu, Hou L, Guo T, Zhao H, Xing M (2023) Polystyrene microplastics promote liver inflammation by inducing the formation of macrophages extracellular traps. J Hazard Mater 452:131236

Zhang K, Hamidian AH, Aleksandra Tubić Yu, Zhang JKH, Fang CW, Lam PKS (2021) Understanding plastic degradation and microplastic formation in the environment: A review. Environ Pollut 274:116554

Zhang W, Sun X, Qi X, Liu X, Zhang Y, Qiao S, Lin H (2022) Di-(2-ethylhexyl) phthalate and microplastics induced neuronal apoptosis through the PI3K/AKT pathway and mitochondrial dysfunction. J Agric Food Chem 70:10771–10781

Zhang Y, Yin K, Dongxu Wang Yu, Wang HL, Zhao H, Xing M (2022) Polystyrene microplastics-induced cardiotoxicity in chickens via the ROS-driven NF-κB-NLRP3-GSDMD and AMPK-PGC-1α axes. Sci Total Environ 840:156727

Zhang Y, Wang S, Olga V, Xue Y, Lv S, Diao X, Zhang Y, Han Q, Zhou H (2022) The potential effects of microplastic pollution on human digestive tract cells. Chemosphere 291:132714

Zhao X, Cornish K, Vodovotz Y (2020) Narrowing the gap for bioplastic use in food packaging: an update. Environ Sci Technol 54:4712–4732

Zhou G, Wang Q, Zhang J, Li Q, Wang Y, Wang M, Huang X (2020) Distribution and characteristics of microplastics in urban waters of seven cities in the Tuojiang River basin, China. Environ Res 189:109893

Zhou Y, Liu X, Wang J (2019) Characterization of microplastics and the association of heavy metals with microplastics in suburban soil of central China. Sci Total Environ 694:133798

Zong W-X, Rabinowitz JD, White E (2016) Mitochondria and cancer. Mol Cell 61:667–676

Assessment of Haematological and Respiratory Health Risk from Air Pollution

Mallika Chowdhury, Buddhadev Ghosh, Rumi Rabha, and Pratap Kumar Padhy

1 Introduction

Air pollution is a primary environmental concern troubling both developing and developed countries. Around 7 million people died in 2012 due to household and ambient air pollution (WHO, 2014). Approximately one in eight deaths globally are caused by air pollution (WHO, 2014). Based on estimates from the Global Burden of Diseases, Injuries, and Risk Factors Study (GBD, 2016), ambient $PM_{2.5}$ is expected to cause 105.7 million disability-adjusted life years (DALY) and 4.1 million premature deaths globally. At present, urbanization and industrialization, along with increasing population are causing serious environmental challenges where automobiles are the primary culprit of ambient air pollution. The world's public health is seriously threatened by both indoor and outdoor air pollution. Utilizing biomass fuels and technology results in higher indoor air pollution levels and a variety of harmful pollutants, such as minute soot particles that penetrate deeply into the lungs (WHO, 2021). More than 2.6 billion people continue to cook with kerosene, which contaminates the air and solid biomass i.e., wood, charcoal, coal, agricultural rubbishes, and dung over open flames and inefficient stoves (WHO, 2021). In comparison with outdoor, several pollutants can be many times higher in indoor and can pose a biological severe impact at low concentrations because of prolonged exposure periods.

Every year, illnesses brought on by domestic air pollution from the inefficient use of kerosene and solid fuels for cooking cause over 3.8 million early deaths. Out of these 3.8 million deaths, pneumonia accounts for 27%, stroke for 18%, ischemic heart disease for 27%, lung cancer for 20%, and chronic obstructive pulmonary disease (COPD) for 8% of deaths (WHO, 2021).

The WHO reports that ambient air pollution kills 4.2 million people globally, with acute lower respiratory infections accounting for 17% of premature deaths,

M. Chowdhury · B. Ghosh · R. Rabha · P. K. Padhy (✉)
Department of Environmental Studies, Visva-Bharati, Santiniketan, West Bengal, India
e-mail: pkpadhy@visva-bharati.ac.in

P. K. Padhy et al. (eds.), *Airborne Particulate Matter*,
https://doi.org/10.1007/978-3-031-84408-9_7

lung cancer for 29%, stroke for 24%, ischemic heart disease for 25%, and chronic obstructive pulmonary diseases for 43% (Fig. 1).

Asthma is one of the respiratory disorders that are quickly increasing in urban areas every day (De Meij et al., 2009); it has 330 million cases worldwide. The reason for 3,83,000 deaths was associated with asthma reported by WHO in 2016 (Fazlollahi et al., 2018; WHO, 2020a, b). Extreme levels of particulate matter (PM) have been reported in a number of countries, mostly by burning of coal, natural gas, and fossil fuels. According to Visani et al. (2016), PMs are intricate combinations of different-size airborne particles, origin, and elemental composition. They are divided into three categories, viz. ultrafine particulate matter (UFP; size < 100 nm), fine particulate matter ($PM_{2.5}$; size < 2.5μm), and coarse particulate matter (PM_{10}; diameter range 2.5–10 μm) (Robertson and Miller, 2018; USEPA, 2019; Wardoyo et al., 2019). While PM_{10} remains in the upper respiratory tracts and nasal cavities after entering the human body, $PM_{2.5}$ is easily able to pass through the alveoli and enter the circulation (Visani et al., 2016). According to Mannucci et al. (2015) and Cohen et al. (2017), fine PM can directly display health issues like respiratory (reactive airway, lung cancer and chronic obstructive pulmonary illnesses), cardiovascular, and allergy diseases. In addition to affecting the respiratory system, PM can affect platelet, white blood, and red blood cell activities. As a result, certain patients may develop haematological disorders. In addition to $PM_{2.5}$ exposure, several risk factors work in concert to cause these disorders. The main target are women and children, as they are more likely to be exposed to PM and have a higher risk of acquiring respiratory diseases (8–20%) and cardiovascular disorders (18–27%) (WHO, 2012). Mohammedi et al. (2019) have reported that gaseous air pollutants, including sulphur dioxide (SO_2), nitrogen dioxide (NO_2), ozone (O_3), BTEX and other organic compounds, are also responsible for the rising prevalence of asthma and other respiratory illnesses. There is a higher chance of developing asthma when breathing in air pollutants such SO_2, NO_2, O_3, CO, PM_{10} and $PM_{2.5}$ (Ding et al., 2017; Buteau et al., 2018; Zhao et al., 2020). Numerous epidemiological studies have already discovered correlations between blood count measures, such as WBC (Gondalia et al., 2020), RBC (Elbarbary et al., 2020; Poursafa et al., 2011), or platelet count (Hou et al., 2020; Zhang et al., 2018), and short- and long-term PM exposure. According to Lassale et al. (2018), hematological factors like red blood cells (RBC), white blood cells (WBC), and platelets (PLAT) are essential in the development of atherosclerosis, one of the underlying illnesses that sets off a cardiovascular incident. Potential mediators of PM's impact on cardiovascular health include alterations in blood count parameters brought by exposure to PM. Numerous animal investigations have demonstrated that acute exposure to outdoor PM appears to create cytokines in the lung that activate the bone marrow to release leucocytes and platelets. These cytokines contribute to the instability of atherosclerotic plaques, increasing the risk of rupture and thrombosis, in conjunction with other proinflammatory and prothrombotic indicators (Mukae et al., 2001; Van Eeden and Hogg, 2002). Reduced RBC synthesis and suppressed RBC maturation may result from the interaction between erythropoietin in the bone marrow and PM-induced cytokines in the lung. According to Lassale et al. (2018) and Van Eeden and Hogg (2002), this will lead to a rise in

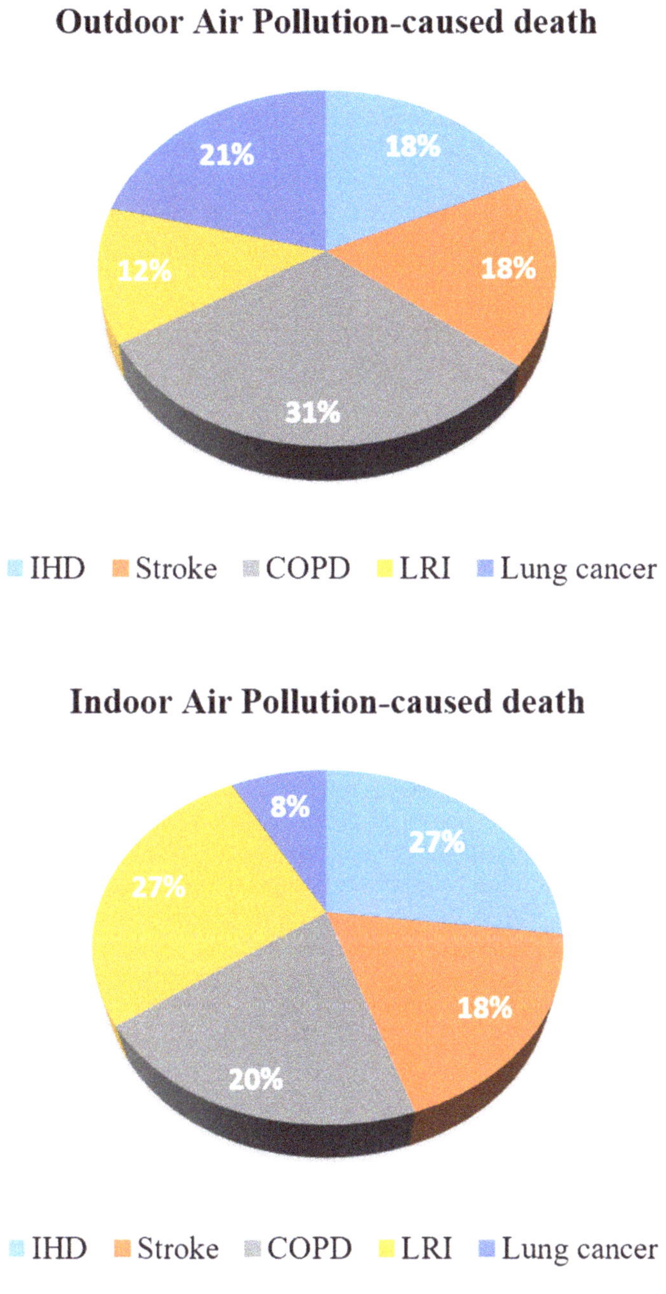

Figure 1: Global deaths breakdown by diseases attributed to outdoor and indoor air pollution in 2021.

immature RBCs and larger red cell distribution width (RDW) values, which might result in anemia and raise the risk of cardiovascular disease under proatherosclеroctic circumstances (Mozos, 2015).

This article focusses on the community health implications of air pollution sources and exposure pathways, particularly with regard to respiratory and haematological risk factors. In order to ensure a better environment for the future, a few problems and solutions that environmental officials and decision-makers would find interesting are also discussed.

2 Sources of Exposure

Large-scale human activities are the primary source of air pollution and discharge the bulk of environmental toxins. These activities involve the usage of combustion engines, power plants, industrial machinery, and automobile emissions (Manisalidis et al., 2020). In addition to these human activities, a number of natural sources, including soil and volcanic eruptions, forest fires, gas stations, field cultivation methods, biomass burning, cleaning and rehabilitation projects, are also influencing our surroundings (Jacobson et al., 2002). The sources are the main factor used to categorize air pollutants. Thus, the four primary sources are: large, local, mobile and natural sources. Some of the major sources of pollution include the metallurgical and other industrial plants, power plants, refineries, petrochemicals, municipal incineration, and the chemical and fertilizer sectors. The inadequate burning of biomass fuels, dry cleaners, printing presses, gas stations, and ambient cigarette smoke are some of the contributors of indoor air pollution. Different vehicles including cars, trains and airplanes, are examples of mobile sources and physical disasters including dust storms, volcano erosion, forest fires, and agricultural burning are natural origins examples.

3 Exposure Assessment Tools by Routes

A path via which a pollutant enters a person or population following contact is called exposure route (IPCS, 2004). Ingestion, cutaneous, or inhalation are the three basic routes of exposure.

3.1 Inhalation

Inhalation route is the primary pathway that plays a significant role in exposure to pollutants in the human body through nose (Fig. 2). Exposure to airborne pollutants arises from inhaling particulate matter (dust), vapours (volatile or semi-volatile

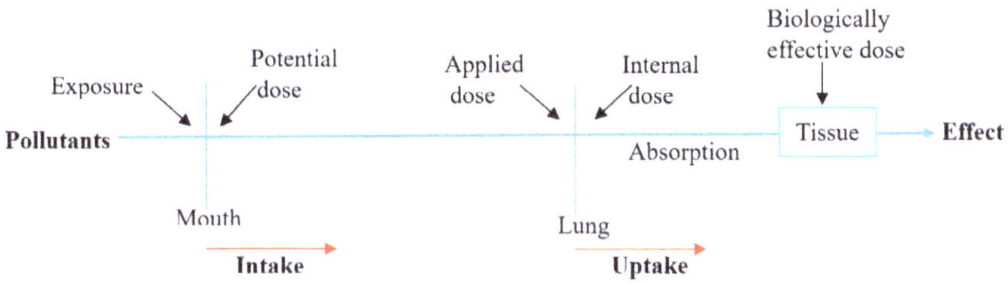

Figure 2: Inhalation route: Exposure and dose (USEPA, 1992a, b, c).

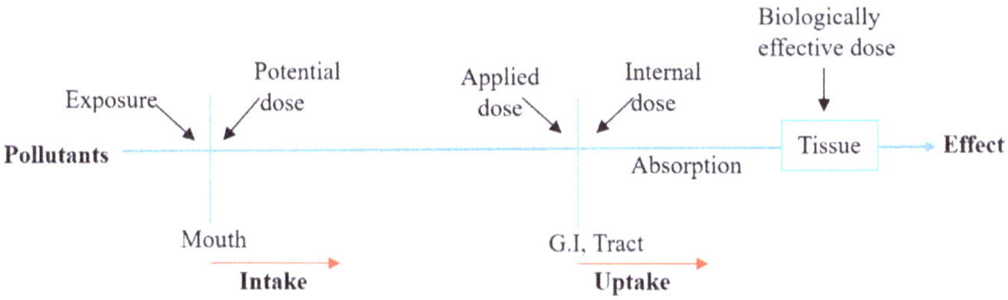

Figure 3: Ingestion route: Exposure and dose (USEPA, 1992a, b, c).

contaminants), or aerosols. People may be exposed by inhalation when engaging in a variety of indoor and outdoor activities. Furthermore, indoor receptors may be subjected to external air pollutants that penetrate into indoor environment.

3.2 *Ingestion*

It is possible to expose oneself to health dangers by intake of contaminated food or liquids. Exposure to non-dietary intake of dirt, dust or chemical residues through hand-to-mouth or object-to-mouth activities can also occur, whether intentional or unintentional (especially for young children (Fig. 3).

3.3 *Dermal*

Dermal exposure can happen when skin is exposed to polluted environmental media, such as water (when bathing, washing, swimming, etc.), sediment (when wading, fishing, etc.), dust or soil from outside (during gardening, building, or engaging in leisure activities), and dust from indoor surfaces (like countertops, carpets, floors or clothes) (Fig. 4).

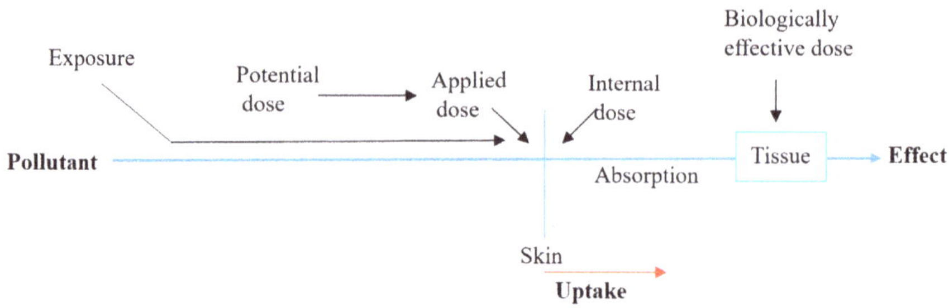

Figure 4: Dermal route: Exposure and dose (USEPA, 1992a, b, c).

4 Air Pollutants

The WHO reports on lead (Pb), carbon monoxide (CO), nitrogen dioxide (NO_2), sulphur dioxide (SO_2), particulate matter (PM), and ground-level ozone (O_3) as the six primary air pollutants. Living things are severely threatened by air pollution. Therefore, since these contaminants are linked to more serious and pervasive issues with human and environmental health, the main focus is on them.

4.1 Particulate Matter (PM)

One common proximate measure for air pollution is PM. This complex combination, which accounts for a large portion of particulate air pollution, is made up of dust or soil particles, metals, organic chemicals and liquid droplets (Anderson et al., 2012). The primary constituents of PM include water, mineral particles, ammonia, sodium chloride, sulfate, nitrates, and black carbon. Both natural and artificial activities contribute to airborne particulate particles. The term PM pollution refers to particles that are 10 μm or smaller, known as PM_{10}, and very fine particles, which are usually 2.5 μm or less (Manisalidis et al., 2020). These coarse particles are mostly the result of uncontrolled combustion, while the fine particles are mostly produced by manufacturing, power generation, agricultural burning, vehicle emissions, and the mechanical breakdown of soil and other crustal materials like coal (Dockery et al., 1994). Inhalation is the main way that exposure to PM affects human health (WHO, 2000).

4.2 Carbon Monoxide

Carbon monoxide is a toxic, colourless, odourless, tasteless, and non-irritating gas, according to the WHO (2010). National ambient air quality standards (NAAQS) for CO are set out by the USEPA Clean Air Act as a criteria air pollutant. It is created when certain fuels, including kerosene, coal, wood, gasoline and natural gas, burn partially. In addition to being produced by these combustion sources during heating and cooking, home air also becomes contaminated with CO from the outside world (Slezakova et al., 2012). It can be dangerous to be exposed to high carbon monoxide levels. Inhalation is the only way to be exposed to carbon monoxide exogenously (WHO, 2010). Inhaling carbon monoxide gas can cause vomiting, headache, weakness, nausea and eventually loss of consciousness (Manisalidis et al., 2020).

4.3 Nitrogen Dioxide

The nitrogen dioxide gas has a strong odour and is reddish-brown in colour. Among secondary pollutants, it is a necessary gas whose impact on human health is well documented (Slezakova et al., 2012). It is a contaminant linked to traffic since automobile engines emit it (Richmont et al., 2017). Elevated levels of nitrogen oxides have been reported as a result of indoor combustion. These are created in ambient air by different combinations of nitrogen and oxygen at high temperatures during combustion; as the temperature rises, more nitric oxide is produced. Since nitric oxide typically emits 90–95% of nitrogen oxides, nitrogen dioxide typically only emits 5–10% (WHO, 2010). The primary indoor sources include coal-burning appliances such as fireplaces, space and water heaters, stoves and ovens; particularly those that are overloaded or badly maintained. Other indoor sources include gas, wood, oil, kerosene and coal-burning equipment (Slezakova et al., 2012; WHO, 2010). In addition to affecting respiratory systems, it irritates eyes, throat and nose (Chen et al., 2007).

4.4 Sulphur Dioxide

Suphur dioxide is a hazardous gas that has a strong odour and is colourless, non-flammable and non-explosive (Slezakova et al., 2012). At concentrations higher than 0.3 ppmv, it has a taste, and at concentrations higher than 0.5 ppmv, a strong odour (WHO, 2010). According to Balakrishnan et al. (2004), the main cause of Suphur dioxide in indoor air is the combustion of coal. According to USEPA (2019), the annual SO_2 standard is 0.03 ppm.

4.5 Lead

Lead is a heavy metal that finds application in several industries. According to the WHO (2010), some gasoline-powered cars, batteries, radiators, wastewater treatment facilities and waste incinerators, emit it. Metals, mining and piston-engine aircraft are the main contributors of airborne lead pollution. Because of lead's detrimental impacts on the environment, lead poisoning poses concern to human health in emerging nations. Ingestion, cutaneous absorption and inhalation are three ways that lead can be exposed.

Among numerous established risk factors for cardiovascular disease is lead. Higher blood lead concentrations in humans have been linked to peripheral artery disease, left ventricular hypertrophy, hypertension, electrocardiographic abnormalities, and cardiovascular disease mortality (Khalil et al., 2009; Aoki et al., 2016). The blood, soft tissues, liver, lungs, bones, cardiovascular, neurological and reproductive systems can all accumulate lead. It may also result in memory loss, pain in the muscles and joints, and loss of focus and concentration (NIH, 2014).

4.6 Ozone

Ozone gas is created when high voltage electric discharge occurs between oxygen molecules (Bezikoglou et al. (2009). It typically originates in the stratosphere, but it can also occur when photochemical haze in the troposphere chains together (Villanyi et al., 2010). Ozone damages DNA more often in epidermal keratinocytes, which reduces cellular activity (McCarthy et al., 2013). Ground-level ozone (GLO) is created by a chemical interaction between nitrogen oxide and volatile organic compounds (VOCs) released by both human activity and natural sources. Exposure through breathing is the most typical way for ozone uptake. According to WHO (2019), ozone has an impact on the tear ducts and the first few layers of skin.

5 Air Pollution and Respiratory Health

The respiratory system is mainly affected by air pollution, and PM is the major contributor to this. Breathing pattern and aerodynamic particle size are the primary determinants of where inhaled particles end up in the respiratory tract. Specifically, during normal nasal breathing, larger particles (>10 μm) are deposited primarily in the extrathoracic part of the respiratory tract, 5–10 μm aerodynamic diameter particles are deposited in the larger conductive airways and 2.5–5 μm are deposited in the smaller conductive airways that are located in close proximity to the fine airways (WHO, 2000).

The PM is responsible for the onset of cardiovascular and cerebrovascular disease due to its mechanisms of systemic inflammation, direct and indirect activation of coagulation, and direct translocation into the systemic circulation. According to Anderson et al. (2012), PM also contributes to respiratory illness and death by inducing inflammation and oxidative stress, which change the anatomy and physiology of the lungs. Fine particulate matter poses a greater health risk to humans (Table 1).

The silent killer CO spreads swiftly to the vital membranes, lungs and capillary arteries (WHO, 2010). Headaches can result from exposure to 300 ppmv of CO for just one hour. Additionally, exposure to 700 ppmv for an hour might be fatal. Inhalation is the main way that NO_2 is exposed (Slezakova et al., 2012). Long-term exposure to high NO_2 mixing ratios damages the lungs and raises the risk of respiratory illnesses. According to WHO (2010) and Slezakova et al. (2012), children suffering from atopy or asthma, babies at risk of developing asthma, and female adults are more vulnerable to illnesses linked to elevated NO_2 exposure. A level of 300 to 800 ppbv lowers lung capacity by 10%, while exposure to more than 80 ppbv increases colds and sore throats (WHO, 2010). Inhalation is the main way that sulphur dioxide is exposed (Slezakova et al., 2012). At mixing ratios higher than 1.5 ppmv, sulphur dioxide can induce bronchiolar constriction in both healthy and asthmatic people, resulting in symptoms such as wheezing, shortness of breath, and respiratory infections (Bernstein et al., 2008; Slezakova et al., 2012). Extended periods of time spent in SO_2 exposure have been linked to respiratory disorders and compromised lung function (WHO, 2010). Some of the diseases caused due to exposure of air pollutants are as follows (Fig. 5).

Table 1: Penetrability based on particle size (Manisalidis et al., 2020)

Particles size	*Level of penetration in the respiratory system of humans*
>11μm	Passages into nostrils and upper respiratory tract
7-11 μm	Passages into nasal cavity
4.7-7 μm	Passages into larynx
3.3-4.7 μm	Passages into trachea-bronchial area
2.1-3.3 μm	Secondary bronchial area passages
1.1-2.1 μm	Terminal bronchial area passages
0.65-1.1 μm	Bronchioles penetrability
0.43-0.65 μm	Alveolar penetrability

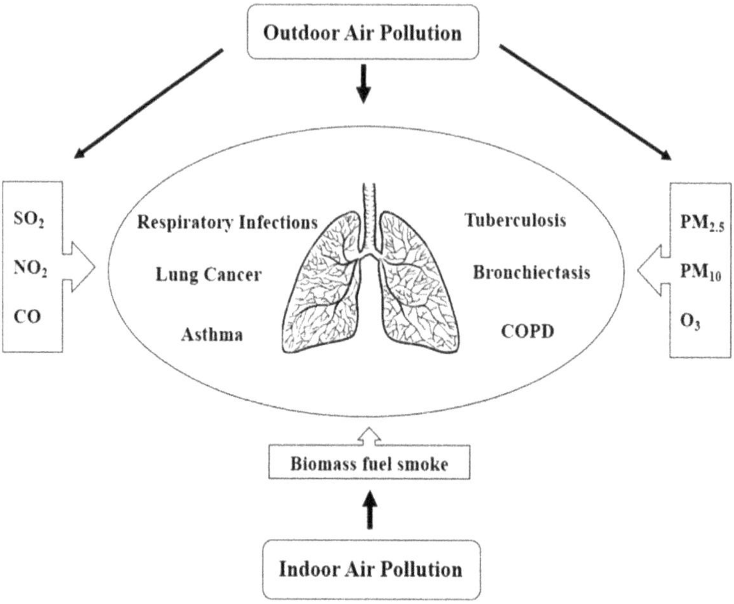

Figure 5: Air pollutants and associated respiratory diseases.

6 Diseases Caused by Air Pollutants

6.1 Asthma

A chronic illness that affects both adults and children is asthma. The muscles around the tiny airways swell and become inflamed, narrowing the airways in the lungs. This results in symptoms of asthma including chest tightness, wheezing, coughing and shortness of breath. These are sporadic symptoms that are frequently worse at night or while exercising. Asthma affected 262 million people and killed 455,000 individuals in 2019, according to WHO estimates (WHO, 2022). High quantities of air pollutants directly cause inflammation in the epithelium and airway neuroreceptors. Pollutant exposure (O_3, NO_2, $PM_{2.5}$) has been linked to oxidative stress (Johannson et al., 2014). Certain pollutants (O_3, NO_2, $PM_{2.5}$) can cause airway inflammation, and O_3 and NO_2 can cause airway hyperresponsiveness (Kravchenko and Lyerly, 2018).

6.2 Lung Cancer

One of the most prevalent cancers is lung cancer, which also has a terrible prognosis. While passive smoking, home radon exposure, and professional environmental carcinogen exposure are all recognized as risk factors for lung cancer, active smoking is still thought to be the biggest danger (Raaschou-Nielsen et al., 2013). The number of lung cancer diagnoses is increasing due to ageing populations, a continuously high smoking prevalence, and continuous exposure to air pollution caused by widespread industrialization (Wang et al. 2019). According to Xing et al. (2019), oxidative stress in the respiratory system is facilitated by $PM_{2.5}$, PM_{10} and O_3. As a result, pulmonary inflammation fosters the mechanisms of carcinogenesis.

6.3 *Chronic Obstructive Pulmonary Disease (COPD)*

The complicated condition known as COPD is marked by a permanent obstruction of the airways, partial reversibility, a progressive course, and an abnormal lung inflammatory response to gases or particles in the air (Singh et al., 2019). There is an association between COPD morbidity and death and ambient air pollution. Research has shown a correlation between a brief rise in air pollution and COPD morbidity (Zhang et al., 2016; Tian et al., 2018). The high prevalence of COPD among non-smokers is associated with domestic air pollution from biomass burning (Gordon et al., 2014).

6.4 *Respiratory Infections*

There has been little investigation on the connection between respiratory diseases and air pollution. Respiratory infections primarily affect children and those with chronic conditions. According to Kurt et al. (2016), people who were exposed to open fire had a 1.91-fold increased risk of acquiring acute respiratory infections (ARI) compared to those who were not.

6.5 *Bronchiectasis*

A persistent and incongruous dilatation of the bronchi is known as bronchiectasis. Numerous people experience recurrent exacerbations due to this chronic respiratory condition. Lung function may subsequently decline as a result of their exacerbations, thus raising mortality (Garcia-Olive et al., 2018). Goeminne et al. (2018) showd that 432 individuals with clinically confirmed bronchiectasis and a diagnosis based on high-resolution computed tomography (HRCT) had a 4.5% and 3.2% greater risk of experiencing an exacerbation that day for every 10 $\mu g/m^3$ rise in PM_{10} and NO_2. In a Badalona observational research, SO_2 was significantly associated with a higher risk of bronchiectasis-related hospitalization (Garcia-Olive et al., 2018).

6.6 Tuberculosis (TB)

As per the WHO-Global Tuberculosis Report 2019, tuberculosis ranks as the tenth most common cause of death worldwide. In 2018, almost 10 million individuals worldwide contacted tuberculosis, the disease caused primarily by one infectious agent (WHO-Global Tuberculosis Report, 2019). The incidence of TB has been linked to socioeconomic risk factors, and it is more predominant in urban areas due to increased exposure to air pollution (Jassal et al., 2013). The incidence of TB and air quality are directly correlated. Rainfall, air pressure and relative humidity all have an indirect effect on TB incidence by lowering the amount of inhalable PM and SO_2 concentrations.

6.7 Air Pollution and Haematological Parameters

Determination of haematological indices such as hemoglobin (Hb), RBC, WBC, hematocrit and platelets are recognized as complete blood count (CBC). The CBC check can help identify acute leukemia in children, subarachnoid hemorrhage, anemia and infections (Llano, 2016; Kilic et al., 2017). It has been shown that exposure to ambient PM causes changes in a number of hematological parameters (Ghio et al., 2003). One of the major contaminants that is dangerous to hematological health is CO. After crossing the tissues of red blood cells, CO penetrates the stroma of these cells, where it combines with hemoglobin to generate carboxyhemoglobin (COHb) (Slezakova et al., 2012). Compared to oxygen, CO has a 200-fold stronger affinity for hemoglobin (Tiwary and Colls, 2009; Slezakova et al., 2012). Hematological markers including factor VII levels, RBC count, hemoglobin, packed cell volume (PCV), and platelets are all altered by exposure to PM_{10} (Seaton et al., 1999). Moreover, increased exposure to PM_{10} and SO_2 levels results in an increase in circulating polymorphonuclear leukocytes (PMN) band cells, which make up 2–6% of the typical circulating WBC population (Athens, 1993; Tan et al., 2000). Early hematological and hemostatic changes may be indicated by an increase in platelet exposure to air pollution (Poursafa and Kelishadi, 2010). A study by Hung et al. (2021) reported that air pollutants, particularly SO_2, NO_2 and CO are linked with WBC amounts and every unit escalation of SO_2, NO_2 and CO is related to WBC count changes of –33.7, –22.3, and 1431.1, respectively which would indicate that air pollution affects systematic inflammation responses. Exposure to air pollutants and their risk is shown in Fig. 6.

In an investigation of the pollutant exposure of 31 healthy volunteers at 5 distinct locations (an underground train station, a farm, two traffic sites and an urban background), Steenhof et al. (2014) found that total WBC, neutrophil, and monocyte counts were positively connected with PM exposure, however there was a negative correlation between NO_2 exposure and lymphocytes and eosinophils. After exposing 15 healthy volunteers to diesel exhaust and ozone, Stiegel et al. (2016) observed

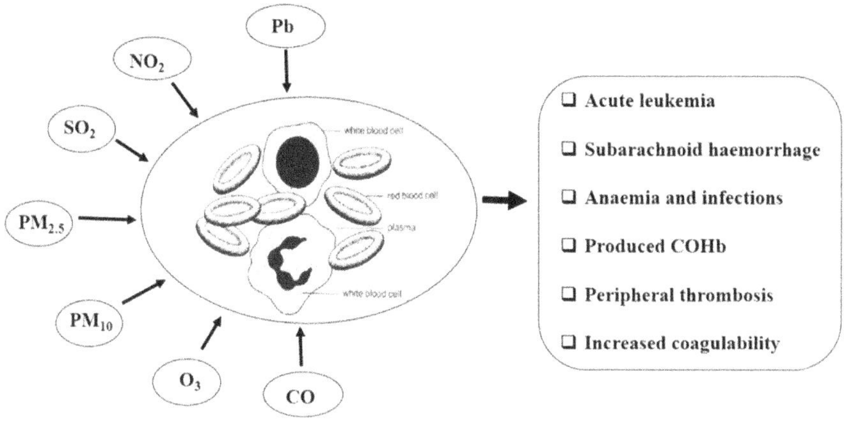

Figure 6: Air pollutants and associated haematological risk.

reductions in monocytes and lymphocytes and proliferations in neutrophils. Nevertheless, it appeared that the overall WBC counts had not changed. Acute phase inflammatory protein fibrinogen, a major cause of amplified viscosity and a known risk factor for venous and arterial thromboses, was shown to be present in higher plasma levels in healthy volunteers who were experimentally exposed to concentrated PM (Ghio et al., 2000). There exists a positive correlation between exposure to traffic-related contaminants, particularly PM and black carbon, and plasma homocysteine, another marker of thrombophilia (Park et al., 2008). A study by Miller et al. (2012) also showed that air pollution exposure was accompanying with the growth in plasma of numerous hemostasis components and also described that over 6 years, for every 10 microgram/m^3 increase in $PM_{2.5}$, exposure to air pollution increased the incidence of cardiovascular events by 24% and cardiovascular-related mortality by 76%.

7 Conclusions

The consequences of air pollution on health are still the main worry. The primary source of the exponential rise in urban air pollution is road traffic, which is a result of global urbanization, industrialization, and vehicle usage. Despite the fact that we are conscious of the impact air pollution has on respiratory and haematological health, we have made an effort to provide accurate information about these relationships. This is contingent upon the kind of pollution, how much of it is present in the ecosystem, and how big it is. To guarantee continuous improvement, it is imperative to measure the air quality on a regular basis. PM and ozone represent the most pervasive health risks for respiratory illnesses among the six criterion pollutants.

Air pollution is a contributing cause to respiratory problems, especially in developing countries where people use coal and biomass fuels for cooking and heating. Exposure is determined by the kind of pollutant, its size, and its concentration in the surrounding environment. The rise in respiratory pathology is amplified by air pollution. It is crucial to continuously monitor the quality of the air in both developed and developing nations in order to guarantee improvement. Even though we are aware of how pollution affects the respiratory system today, we have made an effort to provide current information on this relationship as well as how it impacts haematological indices. Government, business, and the public sector must make decisions at the local, national, and international levels to reduce the impact of air pollution in order to improve health protection for all people on the planet.

References

Anderson JO, Thundiyil JG, Stolbach A (2012) Clearing the air: a review of the effects of particulate matter air pollution on human health. J Med Toxicol 8(2):166–175

Aoki Y, Brody DJ, Flegal KM, Fakhouri TH, Parker JD, Axelrad DA (2016) Blood lead and other metal biomarkers as risk factors for cardiovascular disease mortality. Medicine 95(1):e2223

Athens JW (1993) Granulocytes. In: Lee GR, Bithel TC, Foerster J, Athens JW, Lukens JN (eds) Wintrobe's clinical hematology, 9th edn. Lea and Febiger, Malvern, PA, pp 224t–2303-t

Balakrishnan K, Sambandam S, Ramaswamy P, Mehta S, Smith KR (2004) Exposure assessment for respirable particulates associated with household fuel use in rural districts of Andhra Pradesh, India. J Eposure Sci Environ Epidemiol 14(S1):S14

Bernstein JA, Alexis N, Bacchus H, Bernstein IL, Fritz P, Horner E, ... & Tarlo SM (2008) The health effects of nonindustrial indoor air pollution. J Allergy Clin Immunol 121(3):585–591

Bezirtzoglou E, Alexopoulos A (2009) Ozone history and ecosystems: a goliath from impacts to advance industrial benefits and interests, to environmental and therapeutical strategies. In: Ozone depletion, chemistry and impacts, pp 135–145

Buteau S, Doucet M, Tétreault LF, Gamache P, Fournier M, Brand A, Kosatsky T, Smargiassi A (2018) A population-based birth cohort study of the association between childhood-onset asthma and exposure to industrial air pollutant emissions. Environ Int 121:23–30

Chen TM, Kuschner WG, Gokhale J, Shofer S (2007) Outdoor air pollution: nitrogen dioxide, sulfur dioxide, and carbon monoxide health effects. Am J Med Sci 333(4):249–256

Cohen AJ, Brauer M, Burnett R, Anderson HR, Frostad J, Estep K, Balakrishnan K, Brunekreef B, Dandona L, Dandona R, Feigin V, Freedman G, Hubbell B, Jobling A, Kan H, Knibbs L, Liu Y, Martin R, Morawska L, Pope CA, Shin H, Straif K, Shaddick G, Thomas M, van Dingenen R, van Donkelaar A, Vos T, Murray CJL, Forouzanfar MH (2017) Estimates and 25-year trends of the global burden of disease attributable to ambient air pollution: an analysis of data from the global burden of diseases study 2015. The Lancet 389(10082):1907–1918

De Meij A, Thunis P, Bessagnet B, Cuvelier C (2009) The sensitivity of the CHIMERE model to emissions reduction scenarios on air quality in Northern Italy. Atmos Environ 43(11):1897–1907

Ding L, Zhu D, Peng D, Zhao Y (2017) Air pollution and asthma attacks in children: a case–crossover analysis in the city of Chongqing, China. Environ Pollut 220:348–353

Dockery DW, Pope CA (1994) Acute respiratory effects of particulate air pollution. Annu Rev Public Health 15(1):107–132

Eeden SFV, Hogg JC (2002) Systemic inflammatory response induced by particulate matter air pollution: the importance of bone-marrow stimulation. J Toxicol Environ Health A 65(20):1597–1613

Elbarbary M, Honda T, Morgan G, Guo Y, Guo Y, Kowal P, Negin J (2020) Ambient air pollution exposure association with anaemia prevalence and haemoglobin levels in Chinese older adults. Int J Environ Res Public Health 17(9):3209

Fazlollahi MR, Najmi M, Fallahnezhad M, Sabetkish N, Kazemnejad A, Bidad K, Shokouhi Shoormasti R, Mahloujirad M, Pourpak Z, Moin M (2018) The prevalence of asthma in Iranian adults: the first national survey and the most recent updates. Clin Respir J 12(5):1872–1881

Garcia-Olive I, Stojanovic Z, Radua J, Rodriguez-Pons L, Martinez-Rivera C, Manzano JR (2018) Effect of air pollution on exacerbations of bronchiectasis in Badalona, Spain, 2008–2016. Respiration 96(2):111–116

GBD (2016). Disease and Injury Incidence and Prevalence Collaborators. Global, regional, and national incidence, prevalence, and years lived with disability for 328 diseases and injuries for 195 countries, 1990–2016: a systematic analysis for the global burden of disease study 2016. Lancet 2017 390:1211–1259 [CrossRef]

Ghio AJ, Kim C, Devlin RB (2000) Concentrated ambient air particles induce mild pulmonary inflammation in healthy human volunteers. Am J Respir CritAl Care Med 162(3):981–988

Ghio AJ, Hall A, Bassett MA, Cascio WE, Devlin RB (2003) Exposure to concentrated ambient air particles alters hematologic indices in humans. Inhal Toxicol 15(14):1465–1478

Goeminne PC, Cox B, Finch S, Loebinger MR, Bedi P, Hill AT, Fardon TC, de Hoogh K, Nawrot TS, Chalmers JD (2018) The impact of acute air pollution fluctuations on bronchiectasis pulmonary exacerbation: a case-crossover analysis. Eur Respir J 52(1):1702557

Gondalia R, Holliday KM, Baldassari A, Justice AE, Stewart JD, Liao D, Yanosky JD, Engel SM, Jordahl KM, Bhatti P, Horvath S, Assimes TL, Pankow JS, Demerath EW, Guan W, Fornage M, Bressler J, North KE, Conneely KN, Li Y, Hou L, Baccarelli AA, Whitsel EA (2020). Leukocyte traits and exposure to ambient particulate matter air pollution in the women's health initiative and atherosclerosis risk in communities study. Environ Health Perspect 128(1):017004

Gordon SB, Bruce NG, Grigg J, Hibberd PL, Kurmi OP, Lam K-BH, Mortimer K, Asante KP, Balakrishnan K, Balmes J, Bar-Zeev N, Bates MN, Breysse PN, Buist S, Chen Z, Havens D, Jack D, Jindal S, Kan H, Mehta S, Moschovis P, Naeher L, Patel A, Perez-Padilla R, Pope D, Rylance J, Semple S, Martin WJ II (2014) Respiratory risks from household air pollution in low and middle income countries. Lancet Respir Med 2(10):823–860

Hou J, Duan Y, Liu X, Li R, Tu R, Pan M, Dong X, Mao Z, Huo W, Chen G, Guo Y, Li S, Wang C (2020) Associations of long-term exposure to air pollutants, physical activity and platelet traits of cardiovascular risk in a rural Chinese population. Sci Total Environ 738:140182

Hung SC, Cheng HY, Yang CC, Lin CI, Ho CK, Lee WH, Cheng FJ, Li CJ, Chuang HY (2021) The association of white blood cells and air pollutants—a population-based study. Int J Environ Res Public Health 18(5):2370

Jacobson MZ, Jacobson PMZ (2002) Atmospheric pollution: history, science, and regulation. Cambridge University Press, p 206. https://doi.org/10.1256/wea.243.02

Jassal MS, Bakman I, Jones B (2013) Correlation of ambient pollution levels and heavily-trafficked roadway proximity on the prevalence of smear-positive tuberculosis. Public Health 127(3):268–274

Johannson KA, Vittinghoff E, Lee K, Balmes JR, Ji W, Kaplan GG, Kim DS, Collard HR (2014) Acute exacerbation of idiopathic pulmonary fibrosis associated with air pollution exposure. Eur Respir J 43(4):1124–1131

Khalil N, Wilson JW, Talbott EO, Morrow LA, Hochberg MC, Hillier TA, Muldoon SB, Cummings SR, Cauley JA (2009) Association of blood lead concentrations with mortality in older women: a prospective cohort study. Environ Health 8(1):1–10

Kilic TY, Aksay E, Atilla OD, Sezik S, Camlar M (2017) The diagnostic value of complete blood count parameters in patients with subarachnoid hemorrhage. Turk J Emerg Med 17(4):128–131

Kravchenko J, Lyerly HK (2018) The impact of coal-powered electrical plants and coal ash impoundments on the health of residential communities. N C Med J 79(5):289–300

Kurt OK, Zhang J, Pinkerton KE (2016) Pulmonary health effects of air pollution. Curr Opin Pulm Med 22(2):138

Lassale C, Curtis A, Abete I, van der Schouw Y, Verschuren WM, Lu Y (2018) Elements of the complete blood count associated with cardiovascular disease incidence: findings from the EPIC-NL cohort study. Sci Rep 8(1):1–11

Llano OG (2016) The complete blood count in the early diagnosis of acute leukemia in children. Med Univ 18(73):216–218

Manisalidis I, Stavropoulou E, Stavropoulos A, Bezirtzoglou E (2020) Environmental and health impacts of air pollution: a review. Front Public Health 8:14

Mannucci PM, Harari S, Martinelli I, Franchini M (2015) Effects on health of air pollution: a narrative review. Intern Emerg Med 10(6):657–662

McCarthy JT, Pelle E, Dong K, Brahmbhatt K, Yarosh D, Pernodet N (2013) Effects of ozone in normal human epidermal keratinocytes. Exp Dermatol 22(5):360–361

Miller MR, Shaw CA, Langrish JP (2012) From particles to patients: oxidative stress and the cardiovascular effects of air pollution. Future Cardiol 8(4):577–602

Mohammadi A, Faraji M, Conti GO, Ferrante M, Miri M (2019) Mortality and morbidity due to exposure to particulate matter related to drying Urmia Lake in the NW Iran. Eur J Intern Med 60:e14–e15

Mozos I (2015) Mechanisms linking red blood cell disorders and cardiovascular diseases. BioMed Res Int

Mukae H, Hogg JC, English D, Vincent R, Van Eeden SF (2001) Phagocytosis of particulate air pollutants by human alveolar macrophages stimulates the bone marrow. Am J Physiol-Lung Cell Mol Physiol 279(5):L924–L931

NIH (2014) https://www.nih.gov/news-events/nih-research-matters/low-levels-lead-linked-problems-children

Park SK, O'Neill MS, Vokonas PS, Sparrow D, Spiro III A, Tucker KL, ... Schwartz J (2008) Traffic-related particles are associated with elevated homocysteine: the VA normative aging study. Am J Respir CritAl Care Med 178(3):283–289

Poursafa P, Kelishadi R (2010) Air pollution, platelet activation and atherosclerosis. Inflamm & Allergy-Drug Targets (Former Curr Drug Targets-Inflamm & Allergy) 9(5):387–392

Poursafa P, Kelishadi R, Amini A, Amini A, Amin MM, Lahijanzadeh M, Modaresi M (2011) Association of air pollution and hematologic parameters in children and adolescents. Jornal De Pediatr 87:350–356

Raaschou-Nielsen O, Andersen ZJ, Beelen R, Samoli E, Stafoggia M, Weinmayr G, Hoffmann B, Fischer P, Nieuwenhuijsen MJ, Brunekreef B, Xun WW, Katsouyanni K, DimakopoulouK, Sommar J, Forsberg B, Modig L, Oudin A, Oftedal B, Schwarze PE, Nafstad P, Faire UD, Pedersen NL, Östenson C-G, Fratiglioni L, Penell J, Korek M, Pershagen G, Eriksen KT, Sørensen M, Tjønneland A, Ellermann T, Eeftens M, Peeters PH, Meliefste K, Wang M, Bueno-de-Mesquita B, Key TJ, de Hoogh K, Concin H, Nagel G, Vilier A, Grioni S, Krogh V, Tsai M-Y, Ricceri F, Sacerdote C, Galassi C, Migliore E, Ranzi A, Cesaroni G, Badaloni C, Forastiere F, Tamayo I, Amiano P, Dorronsoro M, Trichopoulou A, Bamia C, Vineis P, Hoek G (2013). Air pollution and lung cancer incidence in 17 European cohorts: prospective analyses from the European Study of Cohorts for Air Pollution Effects (ESCAPE). Lancet Oncol 14(9):813–822.

Richmond-Bryant J, Chris Owen R, Graham S, Snyder M, McDow S, Oakes M, Kimbrough S (2017) Estimation of on-road NO_2 concentrations, NO_2/NO_X ratios, and related roadway gradients from near-road monitoring data. Air Qual Atmos Health 10(5):611–625

Robertson S, Miller MR (2018) Ambient air pollution and thrombosis. Part Fibre Toxicol 15(1):1–16

Seaton A, Soutar A, Crawford V, Elton R, McNerlan S, Cherrie J, Watt M, Agius R, Stout R (1999) Particulate air pollution and the blood. Thorax 11:1027–1032

Singh D, Agusti A, Anzueto A, Barnes PJ, Bourbeau J, Celli BR, Criner GJ, Frith P, Halpin DM, Han M, Varela MV, Martinez F, de Oca MM, Papi A, Pavord ID, Roche N, Sin DD, Stockley R, Vestbo J, Wedzicha JA, Vogelmeier C (2019) Global strategy for the diagnosis, management, and prevention of chronic obstructive lung disease: the GOLD science committee report 2019. Eur Respir J 53(5):1900164

Slezakova K, Morais S, do Carmo Pereira M (2012) Indoor air pollutants: Relevant aspects and health impacts. In: Environmental health-emerging issues and practice. IntechOpen

Steenhof M, Janssen NA, Strak M, Hoek G, Gosens I, Mudway IS, Kelly FJ, Harrison RM, Pieters RH, Cassee FR, Brunekreef B (2014) Air pollution exposure affects circulating white blood cell counts in healthy subjects: the role of particle composition, oxidative potential and gaseous pollutants – the RAPTES project. Inhal Toxicol 26(3):141–165

Stiegel MA, Pleil JD, Sobus JR, Madden MC (2016) Inflammatory cytokines and white blood cell counts response to environmental levels of diesel exhaust and ozone inhalation exposures. PLoS ONE 11(4):e0152458

Tan WC, Qiu D, Liam BL, Ng TP, Lee SH, van Eeden SF, D'Yachkova Y, Hogg JC (2000) The human bone marrow response to acute air pollution caused by forest fires. Am J Respir Crit Care Med 161(4):1213–1217

Tian Y, Xiang X, Juan J, Song J, Cao Y, Huang C, Li M, Hu Y (2018) Short-term effects of ambient fine particulate matter pollution on hospital visits for chronic obstructive pulmonary disease in Beijing, China. Environ Health 17(1):1–8

Tiwary A, Colls J (2009 Air pollution, 3rd edn. Taylor & Francis, London, United Kingdom. ISBN 978-04-1547-933-2

US Environmental Protection Agency. Exposure Assessment Tools by Routes-Dermal (1992). https://www.epa.gov/expobox/exposure-assessment-tools-routes-dermal

US Environmental Protection Agency. Exposure Assessment Tools by Routes-Inhalation (1992). https://www.epa.gov/expobox/exposure-assessment-tools-routes-inhalation

US Environmental Protection Agency. Exposure Assessment Tools by Routes-Ingestion (1992). https://www.epa.gov/expobox/exposure-assessment-tools-routes-ingestion

US Environmental Protection Agency. National Standards Air Quality Criteria for Particulate Matter (2019). https://www.epa.gov/criteria-air-pollutants. Accessed 2 Apr 2021

Villányi V, Turk B, Batic F, Csintalan Z (2010) Ozone pollution and its bioindication. Air Pollut 153:153–176

Visani G, Manti A, Valentini L, Canonico B, Loscocco F, Isidori A, Gabucci E, Gobbi P, Montanari S, Rocchi M, Papa S, Gatti AM (2016) Environmental nanoparticles are significantly over-expressed in acute myeloid leukemia. Leuk Res 50:50–56

Wang N, Mengersen K, Tong S, Kimlin M, Zhou M, Wang L, Yin P, Xu Z, Cheng J, Zhang Y, Hu W (2019) Short-term association between ambient air pollution and lung cancer mortality. Environ Res 179:108748

Wardoyo AY, Juswono UP, Noor JA (2019) How exposure to ultrafine and fine particles of car smoke can alter erythrocyte forms of male mice. Pol J Environ Stud 28(4):2901

WHO (2020a). https://www.who.int/health-topics/air-pollution#tab=tab_1. Accessed 11 Feb 2021

WHO (2020b). https://www.who.int/news-room/fact-sheets/detail/asthma. Accessed 11 Feb 2021

WHO (n.d.). Global tuberculosis report 2019

WHO (2021). 7 million premature deaths annually linked to air pollution. http://www.who.int/mediacentre/news/releases/2014/air-pollution/en/. Accessed 28 Feb 2021

WHO (2022) https://www.who.int/news-room/fact-sheets/detail/asthma

World Health Organization and Inter-Organization Programme for the Sound Management of Chemicals (2004). IPCS risk assessment terminology, vol 1. World Health Organization

World Health Organization (2000) Particulate matter (Chapter 7.3). WHO Regional Publications, Copenhagen, Denmark

World Health Organization (2010) WHO guidelines for indoor air quality: selected pollutants.

World Health Organization (2014) Ambient air pollution database. WHO, Geneva, Switzerland, December 2015. http://www.iarc.fr/en/media-centre/iarcnews/pdf/pr221_E.pdf

World Health Organization (2012). Pollution affects all of us. https://www.who.int/gho/phe/outdoor_air_pollution/exposure/en/. Accessed 2 Apr 2021

Xing DF, Xu CD, Liao XY, Xing TY, Cheng SP, Hu MG, Wang JX (2019) Spatial association between outdoor air pollution and lung cancer incidence in China. BMC Public Health 19(1):1–11

Zhang S, Li G, Tian L, Guo Q, Pan X (2016) Short-term exposure to air pollution and morbidity of COPD and asthma in East Asian area: a systematic review and meta-analysis. Environ Res 148:15–23

Zhang Z, Chan TC, Guo C, Chang LY, Lin C, Chuang YC, Jiang WK, Ho KF, Tam T, Woo KS, Lau AK, Lao XQ (2018) Long-term exposure to ambient particulate matter ($PM_{2.5}$) is associated with platelet counts in adults. Environ Pollut 240:432–439

Zhao S, Liu S, Hou X, Beazley R, Sun Y, Dong S (2020) Evidence of provincial variability in air pollutants-asthma relations in China. J Clean Prod 242:118553

Particulate Air Pollution and Neurotoxicity: A Review

Rupsa Ghosh, Buddhadev Ghosh, Tanmoy Kumar Maharathy Anusree Datta, and Pratap Kumar Padhy

1 Introduction

Air pollution is a major global risk factor that may increase mortality rate (Landrigan et al., 2018; Finch, 2018). Cardiovascular problems, morbidity and mortality arising from arrhythmia, myocardial ischemia, and respiratory illnesses including lung cancer and asthma have all been linked to air pollution (Block and Calderón-Garcidueas, 2009; Mills et al., 2009). Numerous research demonstrates robust link between air pollution and debilitating CNS disorders, including an increased risk of stroke (Chen, 2010), a decline in cognitive function (Calderón-Garcidueas et al., 2008a), Alzheimer's disease (AD), and Parkinson's disease (PD) (Block and Calderón-Garcidueas, 2009). Air pollution induces neurological inflammation, oxidative damage, and neurodegenerative illness, according to research on humans and animals, including mice and canines (Block and Calderón-Garcidueas, 2009). Emissions from diesel exhaust (DE) vehicles are the major contributor of traffic-related air pollution and substantially contribute to global air pollution (Ghio et al., 2012). Few studies have been conducted on the impact of diesel exhaust emissions on human health (Hesterberg et al., 2010; Ma and Ma, 2002). The DE is a substantial contributor to ambient particulate matter, especially fine ($PM_{2.5}$) and ultrafine particulate matter (USEPA, 2002), and consists of more than 40 hazardous air pollutants. The adverse effects of particulate matter can alter the nervous system and also molecular and cellular level. Cruts et al. (2008), documented that DE exposure has a significant effect on the electroencephalogram parameters (delta, theta, alpha, beta) in the human brain (Fig. 1). Mechanically, polluted air may have an impact on the central nervous system (CNS) through a variety of molecular, cellular and inflammatory processes that can cause direct harm to brain tissue or result in

R. Ghosh · B. Ghosh · T. K. M. A. Datta · P. K. Padhy (✉)
Department of Environmental Studies, Siksha Bhavana, (Institute of Science), Visva-Bharati, Santiniketan, Birbhum, West Bengal 731235, India
e-mail: pkpadhy@visva-bharati.ac.in

P. K. Padhy et al. (eds.), *Airborne Particulate Matter*,
https://doi.org/10.1007/978-3-031-84408-9_8

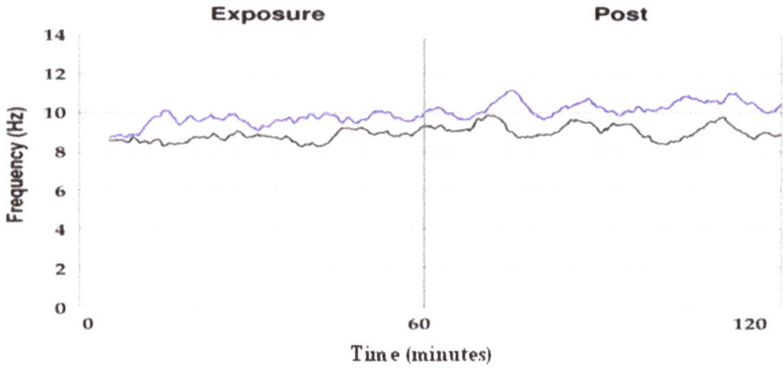

Figure 1: A time history of the median power frequency (MPF) in the left frontal cortex (F3) (Crüts et al., 2008). Values are shown during and after exposure periods for both diesel (blue line) and sham conditions/pure air (black line). The original signal's 5-min moving averages are shown by both lines. During the diesel exposure, MPF levels rise and stay high during the post-exposure period.

neurological diseases. Studies in epidemiology and observational investigations have suggested that exposure to air pollution might cause neurodegenerative illnesses, and this exposure occurs when children are exposed to the pollutants long-term basis (Block and Calderón-Garcidueñas, 2009; Calderón-Garcidueñas and Reed, 2004; Calderón-Garcidueñas et al., 2008a; Calderón-Garcidueñas et al., 2008b). UFPM and $PM_{2.5}$ can enter the system and cause damage to many systems, including the CNS, readily reaching the CNS through the nostril sensory mucosa (Garcia et al., 2015; Lucchini et al., 2012; Peters et al., 2006; Oberdoerster et al., 2004). This article gives an overview of particulate matters and their potential pathways to enter the body. Findings of observational, epidemiological, experimental and clinical studies, which pronounce a relationship between particulate matter and neurological disorders or neurodevelopmental disturbances, have also been discussed.

2 Composition of Air Pollution

Air pollution contains numerous hazardous substances (pollutants) that can induce toxicity if exposed for a prolonged period. Air pollution primarily comprises particulate matter and gaseous substances (sulfur oxides, nitrogen oxides, tropospheric ozone, carbon monoxide), organic compounds (polycyclic aromatic hydrocarbons), and toxic metals (lead, cadmium, chromium, nickel, copper and manganese) and it can be found in both outdoor and indoor atmosphere (Block and Calderón-Garcidueñas, 2009; Craig et al., 2008). Volatile organic compounds and oxides of nitrogen, in the presence of sunlight, primarily act as a precursor of tropospheric ozone formation, which appears to be the most harmful among all the components of air pollution along with PM and other pollutants. Among these, PM, present in the atmosphere as a mixture of liquid tiny particles and solid fragments floating in the air, is mostly responsible for the formation of neurological illnesses (Block and Calderón-Garcidueas, 2009). The main causes of PM_{10} and $PM_{2.5}$ include road dust, tire wear emissions, agricultural dust, wood burning, building and demolition operations and mining activities. According to Craig et al. (2008), five particles often come from power plants, wildfires, residential fuel burning, metal processing facilities, oil refineries, and tailpipe emissions. Block and Calderón-Garcidueas (2009) state that $PM_{2.5}$ can contain both organic and inorganic compounds, such as nitrates, sulphates, carbon, hydrogen ions, ammonium and metals. Diesel exhaust particles (DEP), however, make up the majority of delicate ambient PM.

Particulate matter is usually characterized by an aerodynamic diameter that is directly proportional to its biological effects. The coarse particles are defined as aerodynamic (diameters from >2.5 to 10 μm and <2.5 μm) called fine particles. For instance, PM_{10} (<10 μm) and $PM_{2.5}$ (<2.5 μm) can enter our respiratory system. Fine particles can reach the lungs' inner airways, and coarse particles typically get stuck in the upper airways. Other PM like UFPM and nano-sized particles (<0.1 μm) can also pose threats to physiological systems (Craig et al., 2008; Mazzoli-Rocha et al., 2010). The majority of ultrafine PM comes from combustion, which can occur in power plants, internal combustion engines, incinerators and other thermal degradation sources. UFPMs contain polycyclic aromatic hydrocarbons, soluble organic compounds, and oxidized transition metals (Scheepers and Bos, 1992). UFPMs have been shown to interrupt cellular mechanisms like inhibition of phagocytosis and stimulation of inflammatory responses (Terzano et al., 2010). Concentrations of ambient UFP are reported to be directly inter related with a mortality rate (Wichmann and Peters 2000). UFPs pose a serious risk since they cannot be filtered out during their transit via the nose and bronchioles. In addition, they have the ability to reach deep into the lungs, from whence they may eventually enter the circulatory system and spread throughout the body.

3 Exposure Pathway of Pollutants into Nervous System

Systemic inflammation is the core mechanism through which PM cause cardiovascular and pulmonary toxicity (Fiordelisi et al., 2017), activation of neuroinflammation, oxidative stress, and microglia. Constant exposure to a certain level of ambient air PM can result in the prolapse of these pollutants directly into the nervous system through various pathways (Fig. 2).

On the other hand, these pollutants may not enter straight to the CNS but could trigger the release of inflammatory compounds, which may produce neuroinflammation eventually neurodegeneration. Once taken up and absorbed into the system, PMs, UFPMs can quickly enter the circulation system and can potentially affect the vascular system itself. Due to prolonged exposure to PM, the systemic inflammation takes place, which triggers the production of inflammatory compounds as a result it induced injury in epithelial and endothelial cells of the respiratory system

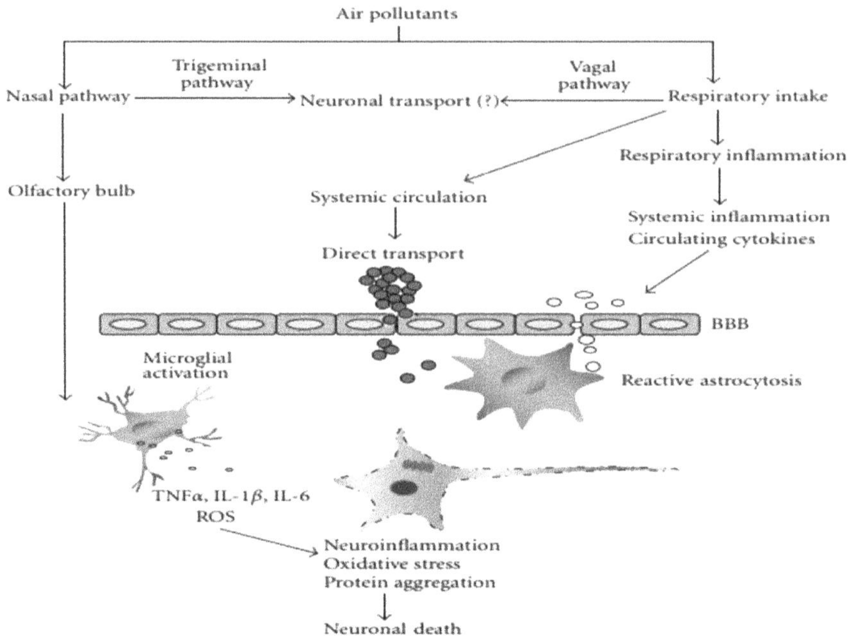

Figure 2: Exposure pathway of pollutants to the CNS and its impact (Genc et al., 2012).

(Cunningham et al., 2009). Tumour necrosis factor (TNF), interleukin-1 (IL-1), and interleukin-6 (IL-6) are examples of proinflammatory cytokines that are produced as a result of systemic inflammation and that have expression-induced receptors in the blood vessels of the brain (Block and Calderón-Garcidueñas, 2009; Teeling and Perry, 2009). All the abovementioned cytokines are known for inducing neuroinflammation, neurotoxicity (Qin et al., 2007; Ling et al., 2006; Perry 2007), and cerebrovascular damage (Manousakis et al., 2009).

These cytokines either excite cerebral endothelial cells, compromise the blood-brain barrier (BBB), or start cell signalling cascades that activate pathways regulated by nuclear factor kappa-B (NF-B) and mitogen-activated protein kinase (MAP). Transferring mast cells and inflammatory cells with cytokine receptors to the damaged location can then occur after BBB disruption (Calderón-Garcidueñas et al., 2008a). Moreover, the circulating cytokines released by endothelial cells or inflamed peripheral organs could penetrate the brain by active transport and diffusion, thus deteriorating the conditions synergistically (Dantzer et al., 2008).

3.1 *Particulate Matter Induced Neurotoxicity in Humans*

According to all the evidence supporting the contribution of PM-induced neurotoxicity, it has come to the point where different effects of FPM and UFPM on the human nervous system have to be scrutinized. Some studies showed adverse effects of air pollutants, specifically PM, on cognitive deterioration and enhanced ageing of the brain. For example, a study on numerous cross-sections on the brain stated decreased cognitive performance amongst aged people residing in higher-level exposures of $PM_{2.5}$ neighbourhoods (Ailshire and Clarke, 2015; Gatto et al., 2014; Ailshire and Crimmins, 2014) or else near major highways (Wellenius et al., 2012).

4 Effect of PM on Brain Structure

Women's Health Initiative Memory Study (WHIMS) enrolled 1403 aged women (65-80 years) in 1996-1998 to observe the effect of $PM_{2.5}$ on brain structure and volume. Cumulative $PM_{2.5}$ exposure was estimated by the use of a spatiotemporal model in the participant's residential areas. At that time, MRI showed 44 patients were diagnosed with incident mild cognitive impairment and 16 others with probable dementia. Without making solid assumptions about the environmental process's distributions, the Bayesian Maximum Entropy (BME)-based environmental spatiotemporal interdependence data was molded to estimate covariance and mean trends of the air pollution fields over time (Chen et al., 2015; Christakos et al., 2001). The data showed that annual exposures of BME were highly associated with the yearly data of $PM_{2.5}$ concentrations documented at the monitoring spots of the U.S. Environmental Protection Agency (EPA). In 2005-2006, magnetic resonance imaging (MRI) was conducted on all 1403 participants (Jaramillo et al., 2007; Coker et al., 2009) in 14 different sites. This study showed that older women with higher exposure to $PM_{2.5}$ had pointedly smaller white matter. For every interquartile increase of cumulative exposure of $PM_{2.5}$ (3.49 $\mu g/m^3$), the average volume of white matter was 6.23 cm^3 decreased in the total brain and the association areas, 4.47 cm^3 lower, which is identical to 1-2 years of brain aging. $PM_{2.5}$ exposure resulted through the study as smaller white matter volumes in the temporal, frontal lobe and corpus callosum (Table 1) (Chen et al., 2015). An increased risk of stroke has been linked to long-term exposure to particle air pollution (Ljungman and Mittleman, 2014) and impairment of the cognitive system in older adults (Power et al., 2011; Weuve et al., 2012). Long-term FPM exposures have also been connected with fluctuations in cerebral hemodynamics (Wellenius, 2013). From this study, a higher level of $PM_{2.5}$ exposure is the reason behind the decrease of the total cerebral brain. It has been demonstrated that a 2 $\mu g/m^3$ rise in $PM_{2.5}$ may be linked to a –0.32% reduction in overall brain volume (Wilker et al., 2015).

Table 1: Multiple linear deteriorations of volumes of white matter due to FPM ($PM_{2.5}$) exposure

Adjusted analyses	*Total white matter*	*Association brain white matter*	*Frontal white matter*	*Parietal white matter*	*Temporal white matter*	*Corpus callosum*
Model A[a] (N = 1403)	−5.52 ± 1.22 (p < 0.001)	−3.90 ± 1.08 (p < 0.001)	−1.74 ± 0.56 (p = 0.004)	−0.56 ± 0.33 (p = 0.09)	−1.60 ± 0.32 (p < 0.001)	−0.11 ± 0.04 (p = 0.005)
Model B[b] (N = 1310)	−5.79 ± 1.25 (p < 0.001)	−4.13 ± 1.1 (p < 0.001)	−1.83 ± 0.57 (p = 0.003)	−0.68 ± 0.34 (p = 0.04)	−1.62 ± 0.33 (p < 0.001)	−0.11 ± 0.04 (p = 0.006)
Model C[c] (N = 1272)	−6.23 ± 1.28 (p < 0.001)	−4.47 ± 1.12 (p < 0.001)	−2.04 ± 0.59 (p = 0.001)	−0.73 ± 0.34 (p = 0.03)	−1.70 ± 0.33 (p < 0.001)	−0.12 ± 0.04 (p = 0.005)

[a]: Adjusted for the intracranial volume, age, race, and geographical region.
[b]: Adjusted for Model A covariates along with socioeconomic status (education, employment, and income status), lifestyle factors (alcohol use, smoking, and physical activities), depressive symptoms, hormone replacement therapy, and body mass index (BMI).
[c]: Adjusted for Model B covariates along with risk factors (diabetes mellitus, hypertension, and hypercholesterolemia), conventional cardiovascular disease (CVD), and CVD histories (Chen et al., 2015).

5 Role of FPM in Alterations of Cerebrovascular Hemodynamics

Studies have shown that when $PM_{2.5}$ levels are momentarily increased, the risk of acute cerebrovascular events increases and disturbance in cerebrovascular hemodynamics function is associated with stroke in individuals (Silvestrini et al., 2000; Troisi et al., 2012). In one study, 482 individuals above the age of 65 years received transcranial Doppler ultrasonography (TCD) for a brief period of time between 2005 and 2008 to quantify the cerebrovascular hemodynamics noninvasively at rest and during stimulating situations. Regular measurements of the PM concentration were also made in the participants' residence areas. A larger cerebral vascular resistance and a slower resting blood flow velocity were associated with an interquartile range increase in mean $PM_{2.5}$ levels over 28 days. The percentiles of $PM_{2.5}$ levels were compared with this data, which was averaged over 28 days, and it was found that there was an 8.6% higher cerebrovascular resistance and a 7.5% lower blood flow velocity (Wellenius et al., 2013).

5.1 *FPM as Stroke Inducing Factor*

One of the most predominant CNS disorders is stroke which can be produced by fine particulate matter ($PM_{2.5}$) exposure. The relationship between stroke and air pollution was first informed after the Great London fog (Mateen and Brook, 2011), but identical results were found from different topographical regions that include Japan, Canada, Italy, USA, Sweden, UK, Taiwan, France and Korea (Henrotin et al., 2007; Hong et al., 2002; Lisabeth et al., 2008; Maheswaran et al., 2005; Miller et al., 2007; O'Donnell et al., 2011; Oudin et al., 2010; Tsai et al., 2003; Vidale et al., 2010; Yorifuji et al., 2011).

Remarkably, a study from Canada presented that only a definite subgroup of diabetes mellitus suffering patients was at high risk for ischemic stroke due to $PM_{2.5}$ exposure (O'Donnell et al., 2011). Age and sex may also differ the effect of air pollution in the production of ischemic stroke. Mostly women and elderly people appear to be sensitive to air pollutants (Hong et al., 2002). Studies reveal that the risk of stroke increases due to both long- and short-term exposure to outdoor $PM_{2.5}$ (Miller et al., 2007; Johnson et al., 2010), even though the stroke risk are less prominent due to long-term exposure (Johnson et al., 2010).

6 Alzheimer's Disease and Air Pollution

Among all the types of dementia, Alzheimer's disease (AD) is the prime type present in individuals over the age of 65 years (Prince et al., 2013), and is one of the massive health care encounters of the 21st century (Scheltens et al., 2016). Several studies performed in Taiwan, Germany, Canada and the USA proved that long-term exposure to PM_{10}, $PM_{2.5}$, O_3, NO_2 is responsible for the initiation of Alzheimer's in individuals (Costa et al., 2020). An AD patient's brain displays diffuse hippocampal and cortical atrophy, development and accumulation of tau proteins, and abnormally folded amyloid beta proteins (Aβ) in amyloid plaques (Fig. 3) and neuronal knots, that defined as neuropathological trademarks of AD (Khan and Bloom, 2016; Selkoe and Hardy, 2016).

Amyloid precursor protein (APP) is degraded by secretases (β and γ secretases) to generate a soluble Aβ polypeptide Aβ42 or Aβ40 with the former being the major form found in amyloid plaques (Fig. 4) (Pressman and Rabinovici, 2014).

Increased AD markers were discovered in several brain areas of autopsy brain samples from toddlers to young adults who lived in the heavily polluted urban area. In a previous study, higher levels of cyclooxygenase2 (COX2), a neuroinflammation indicator triggered by proinflammatory cytokines and raised in AD patients, were discovered in the hippocampus and frontal cortex of people exposed to air pollution (Calderón-Garcidueñas et al., 2004). A particular study on a group of children and young adults with an average age of 13 years was conducted, and it was found

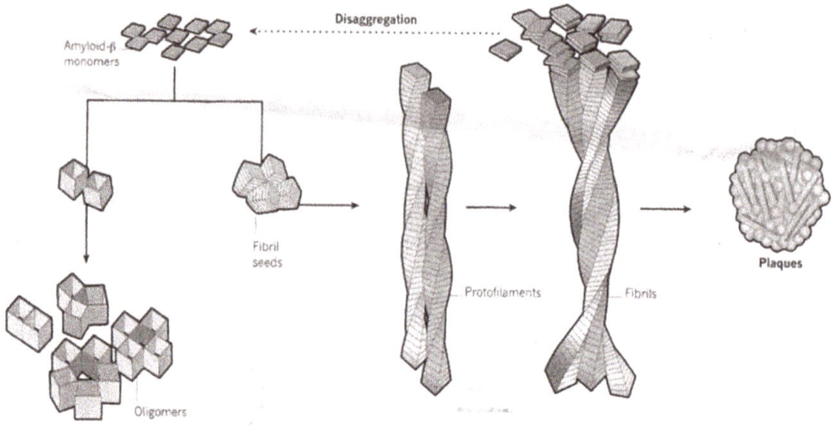

Figure 3: Accumulation of Aβ and formation of amyloid plaques. (www.benbest.com/lifeext/Alzheimer.html)

Figure 4: Steps involved in neurodegeneration (Alghamdi, 2018).

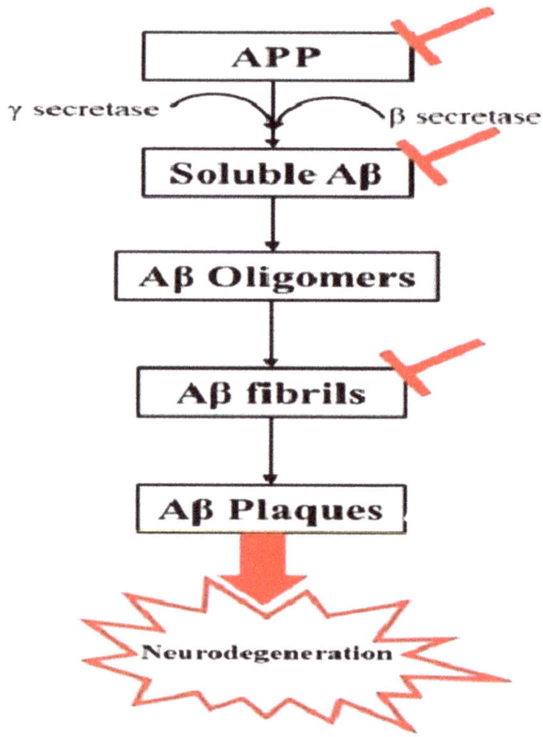

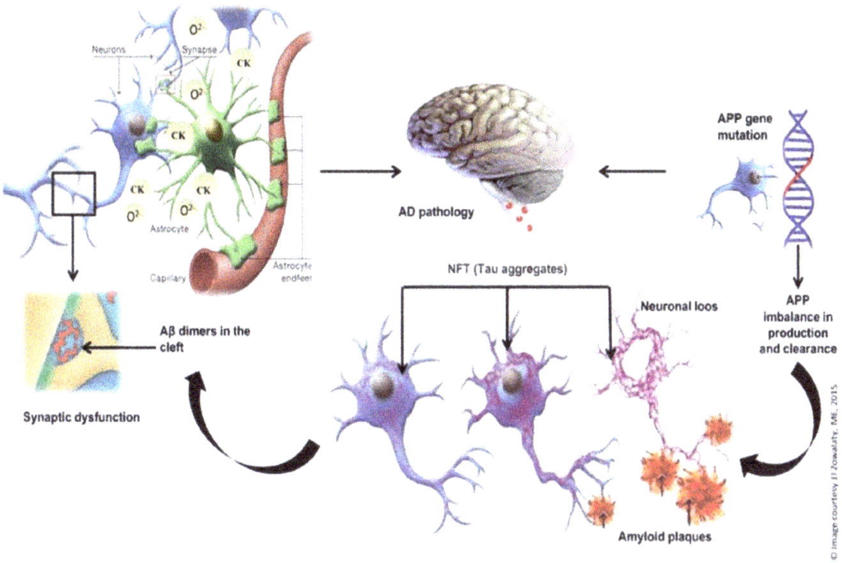

Figure 5: Pathology and molecular mechanisms involved in Alzheimer's disease (Marei et al., 2016).

that 51% of the population had Aβ diffuse plaques and 40% exhibited tau hyperphosphorylation compared to 0% in controls (Calderón-Garcidueñas et al., 2019). A study on older women found long-term exposure to $PM_{2.5}$ is strongly associated with an accelerated decline in episodic memory, immediate recall, and new memory formation. In addition, another study also found $PM_{2.5}$ associated with an increased incidence of AD pattern resemblance (Younan et al., 2020). The Fig. 5 shows the pathology and molecular mechanisms involved in Alzheimer's disease (Marei et al., 2016).

7 Particulate Matter-Induced Autism Spectrum Disorder

According to Levy et al. (2010), autism spectrum disorder (ASD) is a neurodevelopmental syndrome characterised by a substantial decline in communication and social abilities as well as the emergence of stereotyped behavioural traits. Typically, ASD refers to autism as well as several other illnesses that are similar. Autism symptoms commonly coexist with impairments in learning, cognitive functioning, sensory processing and attention before the age of 3 years.

The ASD may be caused by the complex interactions between several genes that increase vulnerability and other environmental conditions (Hertz-Picciotto et al., 2018a, b). In addition to environmental factors, a number of chemicals, such as air pollution, other drugs, and medicines, may have an impact on ASD. Pesticides (organophosphates), polychlorinated biphenyls (PCBs), phthalates, polybrominated diphenyl ethers (PBDEs), and CNS drugs like valproic acid and antidepressants are all associated with ASD (Hertz-Picciotto et al., 2018a, b; Sharma et al., 2018).

The brains of children with ASD have many morphological abnormalities (Gilbert & Man, 2017; Kulesza et al., 2011; Levy et al., 2010; Stoner et al., 2014; Wegiel et al., 2010), and certain neurotransmitter systems (such as glutamate, GABA and serotonin) have changes (Lam et al., 2006; Coghlan et al., 2012). Reelin, a glycoprotein that regulates neuronal migration in the developing central nervous system, has been linked strongly to autism (Fatemi et al., 2005; Folsom & Fatemi, 2013). Increasing evidence from studies designates that children with ASD have elevated levels of oxidative stress (Frustaci et al., 2012; Napoli et al., 2013; Rose et al., 2012), as well as neuroinflammation, microglial activation, and increased systemic inflammation (Chez et al., 2007; Morgan et al., 2010; Ashwood et al., 2011; El-Ansary & Al-Ayadhi, 2012; Tetreault et al., 2012; Theoharides et al., 2013; Suzuki et al., 2013; Depino, 2013). Another study also found that $PM_{2.5}$ and PM_{10} exposure during the early 3 years of life shows elevated risk of ASD (Chen et al., 2018). Figure 6 illustrates the possible molecular mechanisms involved with autism spectrum disorder in newborns (Kamel & Alamodi, 2016).

A study from the "Nurses' Health Study II", reported that perinatal exposure to diesel exhaust was suggestively associated with ASD, predominantly in boys (Roberts et al., 2013). The other two investigations, conducted in Taiwan and Pennsylvania (Jung et al., 2013; Talbott et al., 2015), likewise found a higher incidence

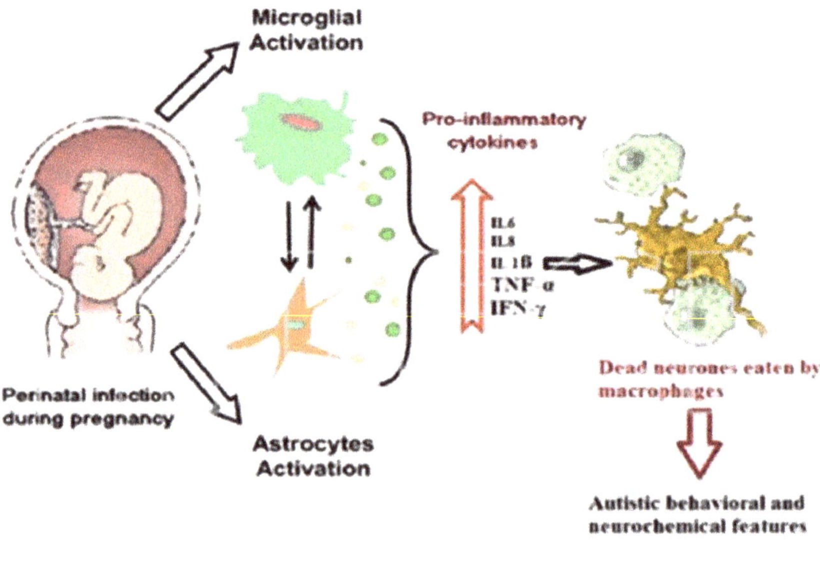

Figure 6: Possible molecular mechanisms involved with autism spectrum disorder in newborns (Kamel & Alamodi, 2016).

of ASD in connection with PM. Some other studies concluded the higher susceptibility towards DE exposure in the third trimester can further initiate ASD in unborn children (Kalkbrenner et al., 2015; Raz et al., 2015).

To support the reports from the human study, experiments on mice were also performed. It showed that prenatal and early life DE exposure of mice is connected with the behavioural aspects observed in humans with ASD. The activation of persisting glial cells, including astrocytes and microglia, as well as ventriculomegaly (lateral ventricular dilation), which only happened in male mice, were demonstrated to occur after postnatal exposure to concentrated ambient UFPMs (Allen et al., 2014a, 2014b). In this latter research, changes in neurotransmitters and cytokines that were sex- and brain region-specific were also discovered (Allen et al., 2014a). It has been demonstrated that prenatal exposure to UFPM during the gestational day (GD0.5 to GD16.5) causes hypermyelination of the cerebellum in only male mice, which is accompanied by an increase in iron content ultrastructural abnormalities, which are also present in people with ASD (Fatemi et al., 2012). The association between PM pollution and ASD is stronger if the exposure occurs in early postnatally or the third trimester of pregnancy, studies have shown (Chen et al., 2018; Kalkbrenner et al., 2015; Raz et al., 2015). This is true even when taking into account differences in the relative rates of brain development between mice and humans. Figure 7 shows the parts of the brain affected by autism spectrum disorder.

8 Conclusions

Air pollution is one of the major global problems and is countable as the chief concern for public health. As a result, governmental, scientific and public communities are highly concerned with the consequences of air pollution. Air pollution is implicated in severe CNS damage and the acceleration of neurodegenerative illnesses, in addition to respiratory and cardiovascular problems, according to mounting data. Different studies concluded that air pollution, or specifically PM pollution attacks the CNS through several molecular and cellular pathways. Effects of PM toxicity in CNS can begin from childhood, before birth, chronic, or may take years to accrue pathology.

Particulate matter pollution specifically promotes neurodegenerative disease, oxidative stress, neuroinflammation, cerebrovascular injury, and microglia activation. In particular, inflammation, persistent activation, and oxidative stress are the reactions of PM pollution on brain capillaries, astroglia and microglia. In certain human and animal test participants, reactions to this toxin have been associated to Alzheimer's disease, an increased risk of stroke, and autism spectrum disorder. Further epidemiological and mechanistic research are urgently needed to identify other severe consequences and the relationship between exposure to air pollution, notably PM pollution, and the emergence of various CNS illnesses.

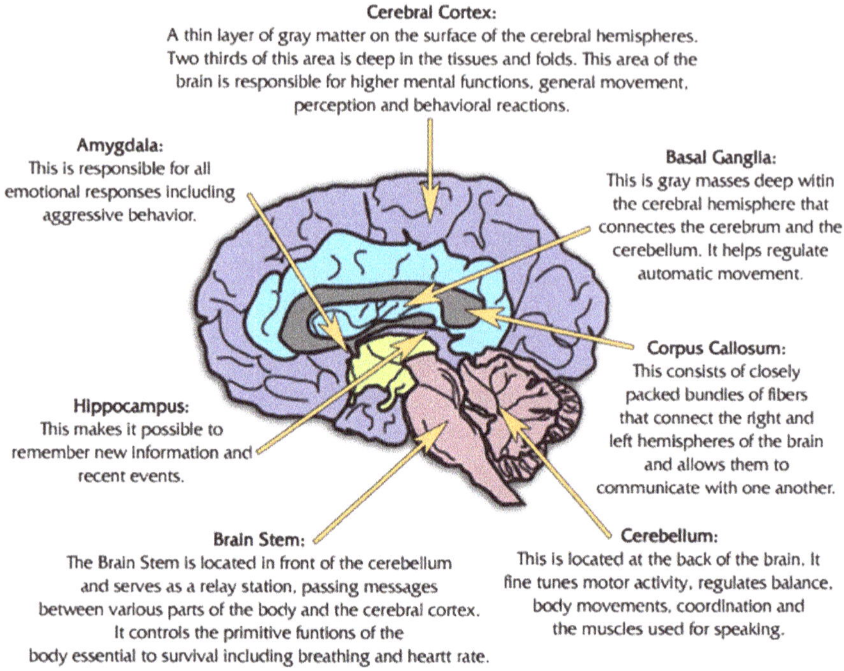

Figure 7: Parts of the brain affected by autism spectrum disorder. (http://hayden26.blogspot.com/2013/10/autism-effects-of-brain-people-with.html)

Acknowledgement B. Ghosh would like to thank the University Grants Commission, New Delhi, for providing Non-NET fellowship during this study.

References

Ailshire JA, Clarke P (2015) Fine particulate matter air pollution and cognitive function among US older adults. J Gerontol B Psychol Sci Soc Sci 70(2):322–328

Ailshire JA, Crimmins EM (2014) Fine particulate matter air pollution and cognitive function among older US adults. Am J Epidemiol 180(4):359–366

Allen JL, Liu X, Pelkowski S, Palmer B, Conrad K, Oberdörster G, Weston D, Mayer-Pröschel M, Cory-Slechta DA (2014a) Early postnatal exposure to ultrafine particulate matter air pollution: persistent ventriculomegaly, neurochemical disruption, and glial activation preferentially in male mice. Environ Health Perspect 122(9):939–945

Allen JL, Liu X, Weston D, Prince L, Oberdörster G, Finkelstein JN, Johnston CJ, Cory-Slechta DA (2014b) Developmental exposure to concentrated ambient ultrafine particulate matter air pollution in mice results in persistent and sex-dependent behavioral neurotoxicity and glial activation. Toxicol Sci 140(1):160–178

Block ML, Calderón-Garcidueñas L (2009) Air pollution: mechanisms of neuroinflammation and CNS disease. Trends Neurosci 32(9):506–516

Calderón-Garcidueñas L, Mora-Tiscareño A, Ontiveros E, Gómez-Garza G, Barragán-Mejía G, Broadway J, Chapman S, Valencia-Salazar G, Jewells V, Maronpot RR, Henríquez-Roldán C, Pérez-Guillé B, Torres-Jardón R, Herrit L, Brooks D, Osnaya-Brizuela N, Monroy ME, González-Maciel A, Reynoso-Robles R, Villarreal-Calderon R, Solt AC, Engle, R.W. (2008a). Air pollution, cognitive deficits and brain abnormalities: a pilot study with children and dogs. Brain Cogn 68(2):117–127

Calderón-Garcidueñas L, Reynoso-Robles R, González-Maciel A (2019) Combustion and friction-derived nanoparticles and industrial-sourced nanoparticles: The culprit of Alzheimer and Parkinson's diseases. Environ Res 176:108574

Calderón-Garcidueñas L, Solt AC, Henríquez-Roldán C, Torres-Jardón R, Nuse B, Herritt L, Villarreal-Calderón R, Osnaya N, Stone I, García, R., Brooks, D.M., González-Maciel A, Reynoso-Robles R, Delgado-Chávez R, Reed W (2008b) Long-term air pollution exposure is associated with neuroinflammation, an altered innate immune response, disruption of the blood-brain barrier, ultrafine particulate deposition, and accumulation of amyloid β-42 and α-synuclein in children and young adults. Toxicol Pathol 36(2):289–310

Calderón-Garcidueñas L, Reed W, Maronpot RR, Henríquez-Roldán C, Delgado-Chavez R, Calderon-Garciduenas AN, Dragustinovis I, Franco-Lira M, Aragón-Flores M, Solt AC, Altenburg M, Torres-Jordón R, Swenberg JA (2004) Brain inflammation and Alzheimer's-like pathology in individuals exposed to severe air pollution. Toxicol Pathol 32(6):650–658

Chen G, Jin Z, Li S, Jin X, Tong S, Liu S, Xu W, Bai Z, Deng F, Zheng X, Guo Y (2018) Early life exposure to particulate matter air pollution (PM1, PM2.5 and PM10) and autism in Shanghai, China: a case-control study. Environ Int 121:1121–1127

Chen JC (2010) Geographic determinants of stroke mortality: role of ambient air pollution. Stroke 41(5):839–841

Chen JC, Wang X, Wellenius GA, Serre ML, Driscoll I, Casanova R, McArdle JJ, Manson JE, Chui HC, Espeland MA (2015) Ambient air pollution and neurotoxicity on brain structure: evidence from women's health initiative memory study. Ann Neurol 78(3):466–476

Christakos G, Bogaert P, Serre M (2001) Temporal GIS: advanced functions for field-based applications. Springer Science & Business Media

Coghlan S, Horder J, Inkster B, Mendez MA, Murphy DG, Nutt DJ (2012) GABA system dysfunction in autism and related disorders: from synapse to symptoms. Neurosci Biobehav Rev 36(9):2044–2055

Coker LH, Hogan PE, Bryan NR, Kuller LH, Margolis KL, Bettermann K, Wallace RB, Lao Z, Freeman R, Stefanick ML, Shumaker SA (2009) Postmenopausal hormone therapy and subclinical cerebrovascular disease: the WHIMS-MRI study. Neurology 72(2):125–134

Costa LG, Cole TB, Dao K, Chang YC, Coburn J, Garrick JM (2020) Effects of air pollution on the nervous system and its possible role in neurodevelopmental and neurodegenerative disorders. Pharmacol Ther 210:107523

Craig L, Brook JR, Chiotti Q, Croes B, Gower S, Hedley A, Krewski D, Krupnick A, Krzyzanowski M, Moran MD, Pennell W (2008) Air pollution and public health: a guidance document for risk managers. J Toxicol Environ Health A 71(9–10):588–698

Crüts B, van Etten L, Törnqvist H, Blomberg A, Sandström T, Mills NL, Borm PJ (2008) Exposure to diesel exhaust induces changes in EEG in human volunteers. Part Fibre Toxicol 5(1):1–6

Cunningham C, Campion S, Lunnon K, Murray CL, Woods JF, Deacon RM, Rawlins JNP, Perry VH (2009) Systemic inflammation induces acute behavioral and cognitive changes and accelerates neurodegenerative disease. Biol Psychiatry 65(4):304–312

Dantzer R, O'Connor JC, Freund GG, Johnson RW, Kelley KW (2008) From inflammation to sickness and depression: when the immune system subjugates the brain. Nat Rev Neurosci 9(1):46–56

Depino AM (2013) Peripheral and central inflammation in autism spectrum disorders. Mol Cell Neurosci 53:69–76

Fatemi SH, Aldinger KA, Ashwood P, Bauman ML, Blaha CD, Blatt GJ, Chauhan A, Chauhan V, Dager SR, Dickson PE, Estes AM (2012) Consensus paper: pathological role of the cerebellum in autism. Cerebellum 11:777–807

Fatemi SH, Snow AV, Stary JM, Araghi-Niknam M, Reutiman TJ, Lee S, Brooks AI, Pearce DA (2005) Reelin signaling is impaired in autism. Biol Psychiatry 57(7):777–787

Finch CE (2018) The role of global air pollution in aging and disease: reading smoke signals. Academic

Fiordelisi A, Piscitelli P, Trimarco B, Coscioni E, Iaccarino G, Sorriento D (2017) The mechanisms of air pollution and particulate matter in cardiovascular diseases. Heart Fail Rev 22:337–347

Folsom TD, Fatemi SH (2013) The involvement of Reelin in neurodevelopmental disorders. Neuropharmacology 68:122–135

Frustaci A, Neri M, Cesario A, Adams JB, Domenici E, Dalla Bernardina B, Bonassi S (2012) Oxidative stress-related biomarkers in autism: systematic review and meta-analyses. Free Radic Biol Med 52(10):2128–2141

Garcia GJ, Schroeter JD, Kimbell JS (2015) Olfactory deposition of inhaled nanoparticles in humans. Inhal Toxicol 27(8):394–403

Gatto NM, Henderson VW, Hodis HN, John JAS, Lurmann F, Chen JC, Mack WJ (2014) Components of air pollution and cognitive function in middle-aged and older adults in Los Angeles. Neurotoxicology 40:1–7

Genc S, Zadeoglulari Z, Fuss SH, Genc K (2012) The adverse effects of air pollution on the nervous system. J Toxicol

Ghio AJ, Smith CB, Madden MC (2012) Diesel exhaust particles and airway inflammation. Curr Opin Pulm Med 18(2):144–150

Gilbert J, Man HY (2017) Fundamental elements in autism: from neurogenesis and neurite growth to synaptic plasticity. Front Cell Neurosci 11:359

Henrotin JB, Besancenot JP, Bejot Y, Giroud M (2007) Short term effects of ozone air pollution on ischaemic stroke occurrence: a case-crossover analysis from a 10-year population-based study in Dijon, France. Occup Environ Med 64(7):439–445

Hertz-Picciotto I, Sass JB, Engel S, Bennett DH, Bradman A, Eskenazi B, Lanphear B, Whyatt R (2018a) Organophosphate exposures during pregnancy and child neurodevelopment: recommendations for essential policy reforms. PLoS Med 15(10):e1002671

Hertz-Picciotto I, Schmidt RJ, Krakowiak P (2018b) Understanding environmental contributions to autism: causal concepts and the state of science. Autism Res 11(4):554–586

Hesterberg TW, Long CM, Lapin CA, Hamade AK, Valberg PA (2010) Diesel exhaust particulate (DEP) and nanoparticle exposures: what do DEP human clinical studies tell us about potential human health hazards of nanoparticles? Inhal Toxicol 22(8):679–694

Hong YC, Lee JT, Kim H, Ha EH, Schwartz J, Christiani DC (2002) Effects of air pollutants on acute stroke mortality. Environ Health Perspect 110(2):187–191

Jaramillo SA, Felton D, Andrews L, Desiderio L, Hallarn RK, Jackson SD, Coker LH, Robinson JG, Ockene JK, Espeland MA (2007) Enrollment in a brain magnetic resonance study: results from the Women's Health Initiative Memory Study Magnetic Resonance Imaging Study (WHIMS-MRI). Acad Radiol 14(5):603–612

Johnson JY, Rowe BH, Villeneuve PJ (2010) Ecological analysis of long-term exposure to ambient air pollution and the incidence of stroke in Edmonton, Alberta, Canada. Stroke 41(7):1319–1325

Jung CR, Lin YT, Hwang BF (2013) Air pollution and newly diagnostic autism spectrum disorders: a population-based cohort study in Taiwan. PLoS One 8(9):e75510

Kalkbrenner AE, Windham GC, Serre ML, Akita Y, Wang X, Hoffman K, Thayer BP, Daniels JL (2015) Particulate matter exposure, prenatal and postnatal windows of susceptibility, and autism spectrum disorders. Epidemiology 26(1):30–42

Kamel HFM, Al-Amodi HSB (2016) Cancer biomarkers. IntechOpen

Kulesza RJ Jr, Lukose R, Stevens LV (2011) Malformation of the human superior olive in autistic spectrum disorders. Brain Res 1367:360–371

Lam KS, Aman MG, Arnold LE (2006) Neurochemical correlates of autistic disorder: a review of the literature. Res Dev Disabil 27(3):254–289

Landrigan PJ, Fuller R, Acosta NJ, Adeyi O, Arnold R, Baldé AB, Bertollini R, Bose-O'Reilly S, Boufford JI, Breysse PN, Chiles T, Mahidol C, Coll-Seck AM, Cropper ML, Fobil J, Fuster V, Greenstone M, Haines, Hanrahan D, Hunter D, Khare M, Krupnick A, Lanphear B, Lohani B, Martin K, Mathiasen KV, McTeer MA, Murray CJL, Ndahimananjara JD, Perera F, Potočnik J, Preker AS, Ramesh J, Rockström J, Salinas C, Samson LD, Sandilya K, Sly PD, Smith KR, Steiner A, Stewart RB, Suk WA, van Schayck OCP, Yadama GN, Yumkella K, Zhong M (2018) The Lancet Commission on pollution and health. The Lancet 391(10119):462–512

Levy SE, Giarelli E, Lee LC, Schieve LA, Kirby RS, Cunniff C, Nicholas J, Reaven J, Rice CE (2010) Autism spectrum disorder and co-occurring developmental, psychiatric, and medical conditions among children in multiple populations of the United States. J Dev Behav Pediatr 31(4):267–275

Ling Z, Zhu Y, Wai Tong C, Snyder JA, Lipton JW, Carvey PM (2006) Progressive dopamine neuron loss following supra-nigral lipopolysaccharide (LPS) infusion into rats exposed to LPS prenatally. Exp Neurol 199(2):499–512

Lisabeth LD, Escobar JD, Dvonch JT, Sánchez BN, Majersik JJ, Brown DL, Smith MA, Morgenstern LB (2008) Ambient air pollution and risk for ischemic stroke and transient ischemic attack. Ann Neurol 64(1):53–59

Ljungman PL, Mittleman MA (2014) Ambient air pollution and stroke. Stroke 45(12):3734–3741

Lucchini RG, Dorman DC, Elder A, Veronesi B (2012) Neurological impacts from inhalation of pollutants and the nose–brain connection. Neurotoxicology 33(4):838–841

Maheswaran R, Haining RP, Brindley P, Law J, Pearson T, Fryers PR, Wise S, Campbell MJ (2005) Outdoor air pollution and stroke in Sheffield, United Kingdom: a small-area level geographical study. Stroke 36(2):239–243

Ma JY, Ma JK (2002) The dual effect of the particulate and organic components of diesel exhaust particles on the alteration of pulmonary immune/inflammatory responses and metabolic enzymes. J Environ Sci Health C 20(2):117–147

Manousakis G, Jensen MB, Chacon MR, Sattin JA, Levine RL (2009) The interface between stroke and infectious disease: infectious diseases leading to stroke and infections complicating stroke. Curr Neurol Neurosci Rep 9(1):28–34

Marei HE, Althani A, Suhonen J, El Zowalaty ME, Albanna MA, Cenciarelli C, et al (2016) Recent perspective about the amyloid cascade hypothesis and stem cell-based therapy in the treatment of Alzheimer's disease. Front Clin Drug Res Alzheimer Dis 5:144–174

Mateen FJ, Brook RD (2011) Air pollution as an emerging global risk factor for stroke. JAMA 305(12):1240–1241

Miller KA, Siscovick DS, Sheppard L, Shepherd K, Sullivan JH, Anderson GL, Kaufman JD (2007) Long-term exposure to air pollution and incidence of cardiovascular events in women. N Engl J Med 356(5):447–458

Mills NL, Donaldson K, Hadoke PW, Boon NA, MacNee W, Cassee FR, Sandström T, Blomberg A, Newby DE (2009) Adverse cardiovascular effects of air pollution. Nat Clin Pract Cardiovasc Med 6(1):36–44

Morgan JT, Chana G, Pardo CA, Achim C, Semendeferi K, Buckwalter J, Courchesne E, Everall IP (2010) Microglial activation and increased microglial density observed in the dorsolateral prefrontal cortex in autism. Biol Psychiatry 68(4):368–376

Mazzoli-Rocha F, Dos Santos AN, Fernandes S, Ferreira Normando VM, Malm O, Saldiva PHN, Picanço-Diniz DLW, Faffe DS, Zin WA (2010) Pulmonary function and histological impairment in mice after acute exposure to aluminum dust. Inhal Toxicol 22(10):861–867

Napoli E, Wong S, Giulivi C (2013) Evidence of reactive oxygen species-mediated damage to mitochondrial DNA in children with typical autism. Mol Autism 4:1–8

Oberdörster G, Sharp Z, Atudorei V, Elder A, Gelein R, Kreyling W, Cox C (2004) Translocation of inhaled ultrafine particles to the brain. Inhal Toxicol 16(6–7):437–445

Oudin A, Strömberg U, Jakobsson K, Stroh E, Björk J (2010) Estimation of short-term effects of air pollution on stroke hospital admissions in southern Sweden. Neuroepidemiology 34(3):131–142
O'Donnell MJ, Fang J, Mittleman MA, Kapral MK, Wellenius GA, Investigators of the Registry of the Canadian Stroke Network (2011) Fine particulate air pollution (PM2.5) and the risk of acute ischemic stroke. Epidemiology 22(3):422–431
Perry VH (2007) Stress primes microglia to the presence of systemic inflammation: implications for environmental influences on the brain. Brain Behav Immun 21:45–46
Peters A, Veronesi B, Calderón-Garcidueñas L, Gehr P, Chen LC, Geiser M, Reed W, Rothen-Rutishauser B, Schürch S, Schulz H (2006) Translocation and potential neurological effects of fine and ultrafine particles a critical update. Part Fibre Toxicol 3(1):1–13
Power MC, Weisskopf MG, Alexeeff SE, Coull BA, Spiro A III, Schwartz J (2011) Traffic-related air pollution and cognitive function in a cohort of older men. Environ Health Perspect 119(5):682–687
Prince M, Bryce R, Albanese E, Wimo A, Ribeiro W, Ferri CP (2013) The global prevalence of dementia: a systematic review and meta-analysis. Alzheimers Dement 9:63–75
Qin L, Wu X, Block ML, Liu Y, Breese GR, Hong JS, Knapp DJ, Crews FT (2007) Systemic LPS causes chronic neuroinflammation and progressive neurodegeneration. Glia 55:453–462
Raz R, Roberts AL, Lyall K, Hart JE, Just AC, Laden F, Weisskopf MG (2015) Autism spectrum disorder and particulate matter air pollution before, during, and after pregnancy: a nested case–control analysis within the Nurses' Health Study II cohort. Environ Health Perspect 123:264–270
Roberts AL, Lyall K, Hart JE, Laden F, Just AC, Bobb JF, Koenen KC, Ascherio A, Weisskopf MG (2013) Perinatal air pollutant exposures and autism spectrum disorder in the children of Nurses' Health Study II participants. Environ Health Perspect 121:978–984
Rose S, Melnyk S, Pavliv O, Bai S, Nick TG, Frye RE, James SJ (2012) Evidence of oxidative damage and inflammation associated with low glutathione redox status in the autism brain. Transl Psychiatry 2:e134
Scheltens P, Blennow K, Breteler MM, De Strooper B, Frisoni GB, Salloway S, Van der Flier WM (2016) Alzheimer's disease. Lancet 388:505–517
Sharma A, Gokulchandran N, Sane H, Nivins S, Paranjape A, Badhe P (2018) The baseline pattern and age-related developmental metabolic changes in the brain of children with autism as measured on positron emission tomography/computed tomography scan. World J Nucl Med 17:94–101
Silvestrini M, Vernieri F, Pasqualetti P, Matteis M, Passarelli F, Troisi E, Caltagirone C (2000) Impaired cerebral vasoreactivity and risk of stroke in patients with asymptomatic carotid artery stenosis. JAMA 283(16):2122–2127
Stoner R, Chow ML, Boyle MP, Sunkin SM, Mouton PR, Roy S, Wynshaw-Boris A, Colamarino SA, Lein ES, Courchesne E (2014) Patches of disorganization in the neocortex of children with autism. N Engl J Med 370:1209–1219
Suzuki K, Sugihara G, Ouchi Y, Nakamura K, Futatsubashi M, Takebayashi K, Yoshihara Y, Omata K, Matsumoto K, Tsuchiya KJ, Iwata Y (2013) Microglial activation in young adults with autism spectrum disorder. JAMA Psychiatry 70:49–58
Talbott EO, Arena VC, Rager JR, Clougherty JE, Michanowicz DR, Sharma RK, Stacy SL (2015) Fine particulate matter and the risk of autism spectrum disorder. Environ Res 140:414–420
Terzano C, Di Stefano F, Conti V, Graziani E, Petroianni A (2010) Air pollution ultrafine particles: toxicity beyond the lung. Eur Rev Med Pharmacol Sci 14:809–821
Tetreault NA, Hakeem AY, Jiang S, Williams BA, Allman E, Wold BJ, Allman JM (2012) Microglia in the cerebral cortex in autism. J Autism Dev Disord 42:2569–2584
Troisi E, Matteis M, Silvestrini M, Paolucci S, Grasso MG, Pasqualetti P, Vernieri F, Caltagirone C (2012) Altered cerebral vasoregulation predicts the outcome of patients with partial anterior circulation stroke. Eur Neurol 67(4):200–205
Tsai SS, Goggins WB, Chiu HF, Yang CY (2003) Evidence for an association between air pollution and daily stroke admissions in Kaohsiung, Taiwan. Stroke 34:2612–2616

US Environmental Protection Agency (2002) Health assessment document for diesel engine exhaust. National Center for Environmental Assessment OoRaD, p 669

Vidale S, Bonanomi A, Guidotti M, Arnaboldi M, Sterzi R (2010) Air pollution positively correlates with daily stroke admission and in hospital mortality: a study in the urban area of Como, Italy. Neurol Sci 31:179–182

Wegiel J, Kuchna I, Nowicki K, Imaki H, Wegiel J, Marchi E, Ma SY, Chauhan A, Chauhan V, Bobrowicz TW, de Leon M (2010) The neuropathology of autism: defects of neurogenesis and neuronal migration, and dysplastic changes. Acta Neuropathol 119:755–770

Wellenius GA, Boyle LD, Coull BA, Milberg WP, Gryparis A, Schwartz J, Mittleman MA, Lipsitz LA (2012) Residential proximity to nearest major roadway and cognitive function in community-dwelling seniors: results from the MOBILIZE boston study. J Am Geriatr Soc 60(11):2075–2080

Wellenius GA, Boyle LD, Wilker EH, Sorond FA, Coull BA, Koutrakis P, Mittleman MA, Lipsitz LA (2013) Ambient fine particulate matter alters cerebral hemodynamics in the elderly. Stroke 44(6):1532–1536

Weuve J, Puett RC, Schwartz J, Yanosky JD, Laden F, Grodstein F (2012) Exposure to particulate air pollution and cognitive decline in older women. Arch Intern Med 172(3):219–227

Wichmann HE, Peters A (2000) Epidemiological evidence of the effects of ultrafine particle exposure. Philos Trans A Math Phys Eng Sci 358:2751–2769

Wilker EH, Preis SR, Beiser AS, Wolf PA, Au R, Kloog I, Li W, Schwartz J, Koutrakis P, DeCarli C, Seshadri S, Mittleman MA (2015) Long-term exposure to fine particulate matter, residential proximity to major roads and measures of brain structure. Stroke 46(5):1161–1166

Yorifuji T, Kawachi I, Kaneda M, Takao S, Kashima S, Doi H (2011) Diesel vehicle emission and death rates in Tokyo, Japan: a natural experiment. Sci Total Environ 409:3620–3627

Younan D, Petkus AJ, Widaman KF, Wang X, Casanova R, Espeland MA, Gatz M, Henderson VW, Manson JE, Rapp SR, Sachs BC (2020) Particulate matter and episodic memory decline mediated by early neuroanatomic biomarkers of Alzheimer's disease. Brain 143:289–302

Source Apportionment Characterization of Airborne Particulate Matter Induced Toxic Effects and Various Assay of Oxidative Potential Measurement

Buddhadev Ghosh, Sukanta Nayek, and Pratap Kumar Padhy

1 Introduction

Airborne particle matter (PM) and harmful health impacts in people have been connected in a plethora of epidemiological and clinical investigations (Hamanaka and Mutlu, 2018; Jeon et al., 2011; Turner et al., 2011; Cho et al., 2005; Johnson, 2004; Pope et al., 2018, 2009, 2004, 2002; Donaldson et al., 2001). The PM-induced health effect mechanisms are not properly understood as to how it affects on biota (Cho et al., 2005). Numerous studies looked into the possibility that oxidative stress, or the production of reactive oxygen species (ROS), such as superoxide, singlet O_2, H_2O_2 and the extremely reactive OH radicals, was responsible for the adverse health consequences (Fig. 1) (Jeon et al., 2011; Hatfield et al., 2009; Bondareva et al., 2007; Cho et al., 2005; Adler et al., 1999). A study found that PM_{10}-induced DNA damage and apoptosis have a significant role in cancer development (Sánchez-Pérez et al., 2009; Dagher et al., 2007). According to epidemiological research, small particles may be more responsible for harmful effects than coarse particles (Kim et al., 2001). The PM physical properties and chemical composition (heterogeneous mixture) change due to their emission sources and meteorological conditions (WHO, 2006). The size of PMs and their arrangements are directly associated with different toxic impacts on biota. Few studies indicate coarse particles are more effective to induce haemolysis, DNA deterioration (Osornio-Vargas et al., 2011) and proinflammatory (Hetland et al., 2004, 2005; Osornio-Vargas et al., 2003) than fine particles ($PM_{2.5}$). In addition, some study showed that initiation of cell propagation is significantly more potent by

B. Ghosh · P. K. Padhy (✉)
Department of Environmental Studies, Siksha Bhavana (Institute of Science), Visva-Bharati, Santiniketan, Birbhum, West Bengal 731235, India
e-mail: pkpadhy@visva-bharati.ac.in

S. Nayek
Department of Environmental Science, The University of Burdwan Golapbag, Burdwan, West Bengal, India

P. K. Padhy et al. (eds.), *Airborne Particulate Matter*,
https://doi.org/10.1007/978-3-031-84408-9_9

$PM_{2.5}$ (Fang et al., 2017; Ostro et al., 2000). Small particles (< 2.5 μm) can harm cell organelles like mitochondria by penetrating the pulmonary barriers, reaching the deepest region of the lungs, and moving on to other organs or cell membranes (Li et al., 2003). According to recent research, PM with a diameter of up to 1.75 μm can go from the lungs to other organs (Li et al., 2017). According to several studies, PM causes proinflammatory responses in the nostrils, lung tissue, and the circulatory system (Gonzalez-Flecha, 2004; Li et al., 2003; Baulig et al., 2004; Squadrito et al., 2001; Donaldson et al., 2001). The generation of elevated ROS disbalances the redox potential of the cell system (Schafer and Buettner, 2001). ROS triggers the inflammation of organs and cell apoptosis at high concentrations (Nel et al., 2001). According to Cho et al. (2005), redox-reactive chemical species like redox-reactive metals and organic chemicals (transition metals like iron and copper) as well as enzymes that catalyse electron transfer are used in the cell system to reduce oxygen, which results in the production of ROS. Gordon et al. (2013) discovered variations of ROS production in human epithelial and endothelial cell which depends on sampling location, season variations, composition and size of PM. Due to an increase in neutrophils, PM causes ROS generation in the cells and it starts lung inflammation. On an equal-mass basis, ultrafine PM produced higher levels of ROS than coarse and fine PM. Additionally, they discovered that transitional metals including copper (Cu), antimony (Sb), vanadium (V), cobalt (Co), beryllium (Be) and nickel (Ni) were associated with high correlations between ROS generation and PM (Adams et al., 2015). For example, Fang et al. (2017), Zomer et al. (2011), Ayres et al. (2008), Jung et al. (2006), Cho et al. (2005), Mudway et al. (2005), and Venkatachari et al. (2005) developed several assays to measure the oxidative potential of PM samples. Dithiothreitol (DTT) (Cho et al., 2005) and the ascorbic acid (AA) assay (Ayres et al., 2008; Mudway et al., 2005) are the most often used methods for the determination of acellular oxidative potential (OP). The widespread adoption of acellular oxidative potential tests such as glutathione (GSH), electron spin (or paramagnetic) resonance (ESR), AA, and DTT assays. While DTT, AA and GSH assess the rate at which chemical substitutes for cellular reductants (DST) or antioxidants (AA, GSH) are depleted, electron spin resonance (ESR) quantifies the production of hydroxyl radicals.

2 Size Fraction of PM and Sources

2.1 Size

Numerous sizes of PM are present in the atmosphere (Fig. 2) upon which <10 um size PMs are matter of concern. Particles <10 um can reach our lungs through the nasal cavity, whereas, the large size particles (>10 μm) are blocked by nose hair. In general, "coarse (PM_{10})", "fine ($PM_{2.5}$)", and "ultrafine ($PM_{0.1}$)" particles make up the three categories of particulate matter. PM_{10} is a term used to describe airborne particles having an aerodynamic diameter of 10 μm or less, including mould spores, bacteria,

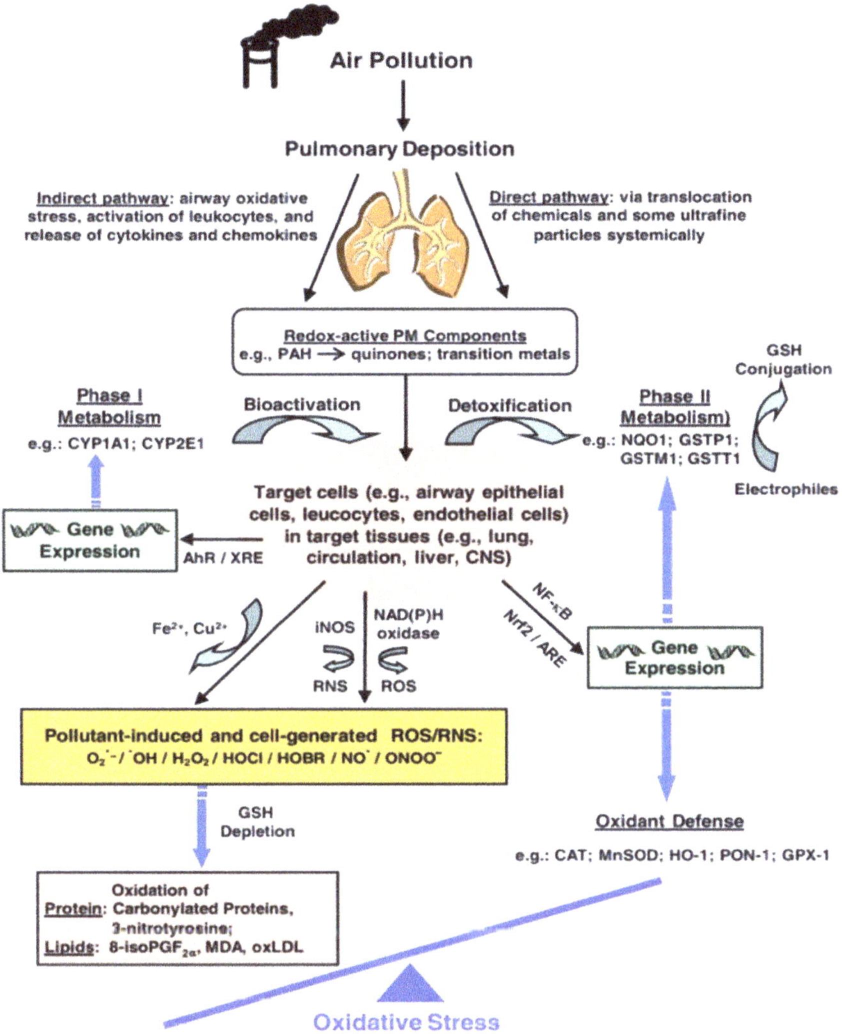

Figure 1: Potential pathways for the production of ROS or RNS during the oxidative stress response to air pollutants (PAH, quinones, PM and PM-bound transition metals) (Delfino et al., 2011).

dust, smoke and airborne virus particles. $PM_{2.5}$ and $PM_{0.1}$ are fine and ultrafine particles, respectively due to the size. Although the danger posed by coarse particles is less than that of fine or ultrafine particles, human health is still very much at risk. These tiny particles have the potential to irritate the eyes, nose, throat and airways in our respiratory system. $PM_{2.5}$ is defined as particles with an aerodynamic diameter equal to 2.5 μm or less such as shoot, black carbon particles, smoke etc. The $PM_{2.5}$ can penetrate the deepest part of our lungs and even in flowing our blood stream. The

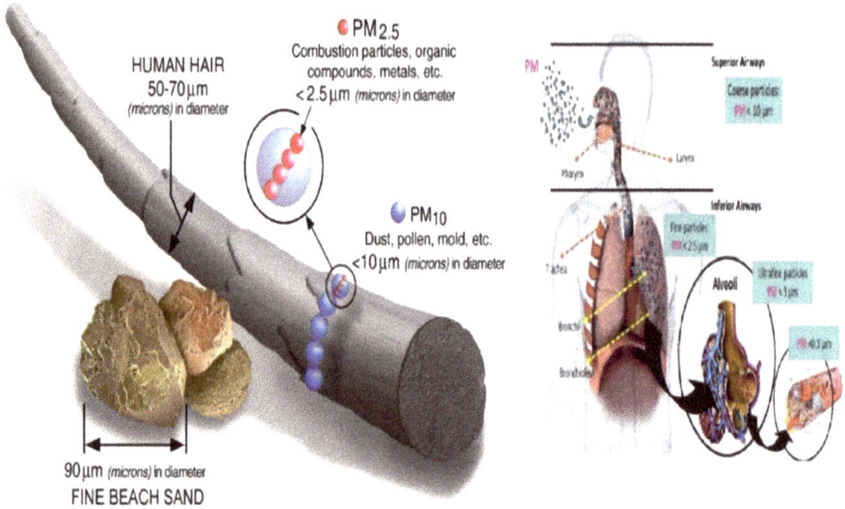

Figure 2: Size fraction of PM and possible pathway to reach into the lungs and blood (ALA, 2020; Castelo, 2017).

ultrafine particles ($PM_{0.1}$) is defined as very tiny particles which is lesser than fine particles with an aerodynamic diameter less than 0.1 µm. This size fraction particles come from similar sources as $PM_{2.5}$. $PM_{0.1}$ is less known for us than $PM_{2.5}$ and PM_{10} because of few research on it. Recent research found that ultrafine particles pose a bigger threat than $PM_{2.5}$ and PM_{10} (Munzel et al., 2018). Additionally, research has shown that $PM_{0.1}$ has a greater propensity for oxidative stress and increased cardiovascular damage.

2.2 *Sources*

The sources of PM depend on various factors, i.e., country's geographical position, meteorological condition and developmental activities such as rapid industrialization (like automobile), urbanization, mining, coal-based power plant set-up, extensive agriculture, build-up road networks, buildings and demolitions. The PM originates from both natural (biogenic) as well as anthropogenic (human) activity. A substantial primary source of PM is generated from various human (anthropogenic) activities, including farming, industrial processes, burning wood and fossil fuels, building and demolition work, and releasing road network dust into the atmosphere (CDC, 2019). In addition, natural sources including windblown and suspended terrestrial dust, oceans and seas salt, volcanic eruption and forest fire or wildfires also contribute to the PM problem. Table 1 shows the major sources and contributions of PM_{10} worldwide. The primary sources of $PM_{2.5}$ can be classified similarly to PM_{10} sources. These sources encompass various sectors such as residential energy consumption (which includes household energy used for cooking and heating, biofuel usage, generators, and small-scale combustion sources), industrial activities (emissions from manufacturing, transformation, chemical production, pulp and paper industry, oil refineries, and fuel/oil production), power generation (emissions from the combustion of fossil fuels for public power consumption), agriculture (mainly gaseous ammonia emissions associated with fertilizer use and domesticated animals), transportation (emissions from vehicles, including air and marine traffic), open fires (forest fires in boreal and temperate regions, peatland fires, savanna fires, grassland and shrubland fires, agricultural waste burning, deforestation and degradation), volcanic sulfur emissions, and mineral dust from arid and semi-arid regions (Weagle et al., 2018). Yu et al. (2013) identified seven types of $PM_{2.5}$ sources in the urban environment of Beijing and their annual contribution which are secondary sulphur (26.5%), vehicle exhaust (17.1%), fossil fuel combustion (16%), road dust (12.7%), soil dust (10.4%), biomass burning (11.2%) and metal processing (6.0%). The source contribution of $PM_{2.5}$ (Qin et al., 2006) in the metropolitan city of New York such as Bronx, Queens, Union and Morris varied. In Bronx, Queens, Union and Morris contribute $PM_{2.5}$ from different sources such as coal burning (1.9%, 1.7%, 0.2% and 0.1%), gas burning (0.0%, 0.0%, 1.3% and 2.4%), oil burning (1.3%, 1.5%, 1.0% and 1.3%), wood burning (6.6%, 16.7%, 17.3% and 19.7%), waste disposal (4.7%, 3.8%, 10.6% and 13.2%), fugitive dust (56.9%, 38.2%, 39.2% and 41.2%) and vehicle emissions (mainly diesel and gasoline-burning) (19.4%, 18.7%, 6.6% and 12.9%). In Birmingham (U.K.), the summer traffic contribution to PM was 23% for PM_{10} and 40% for $PM_{2.5}$, respectively, whereas the winter traffic PM was 41% for $PM_{2.5}$ and 32% for PM_{10} on average (Harrison, 1997). Querol et al. (2001) reported that the highest contribution of $PM_{2.5}$ comes from vehicular emissions and the major sources of PM_{10} and $PM_{2.5}$ were (including marine source) 4% in PM_{10} and negligible in $PM_{2.5}$, the crustal source 26% in PM_{10} and 8% in $PM_{2.5}$, vehicular-secondary 54% in PM_{10} and 73% in $PM_{2.5}$ and unidentified sources 17% in PM_{10} and 17% in $PM_{2.5}$. The sources contributing to $PM_{2.5}$ in Nagpur (India) residential, commercial and industrial regions differ (Pipalatkar et al., 2014). At residential, commercial and industrial locations, vehicle emissions made up the majority of the contributing sources 57, 62 and 65%, followed by secondary inorganic aerosol 16, 12, 16%, biomass burning 15, 11 and 9%, and resuspended dust 7, 10 and 7% respectively.

Table 1: Source distribution of PM_{10} of different city, country (Updated Mukherjee and Agrawal, 2017)

City, Country	*Major sources of PM fraction*
Bern, Switzerland	Local road traffic emissions at the urban roadside location contribute around 30% of the total mass of PM_{10}.
Granada, Spain	At metropolitan locations, the bulk of PM_{10} is made up of mineral matter (43%), secondary inorganic aerosol (SIA) (17%), organic matter, and elemental carbon (31%).
Zaragoza, Spain	Largest source of PM_{10} is soil resuspension (66%) with the next largest sources being industrial and transportation emissions (8%), coal combustion (3%), marine components (3%) and heavy-duty vehicles (1%), as well as unidentified sources (19%).
Puertollano, Spain	PM_{10} has a high percentage of crustal elements (30%).
Zaragoza, Spain	Main causes of PM_{10} levels were soil resuspension (78%) and industrial and transportation emissions (20%).
Palermo, Italy	Road traffic made up over 50% of PM.
Terni, Italy	Road dust from vehicles, secondary/long-range emissions, soil/road dust, and metallurgy/chemical were the main sources found in coarse fractions.
Dresden, Germany	Local sources include crust material, wind-blown dust, road construction, and coal burning are the main contributors of PM_{10} and its related components.
Athens, Greece	Diesel exhaust alone made for around 50% of the bulk of PM_{10}.
Rotterdam, Netherlands	Road traffic emissions, resuspension of road dust, tire/engine wear, burning of biomass and fuel, building activities, and unpaved roads were the main sources of PM_{10} in metropolitan areas.
Dunkirk, France.	Road transport accounts for the largest portion (15%) of PM_{10} emissions, followed by dust resuspension (13%), metallurgical coke plants (12.6%), marine aerosol (12%), crustal dust (11%) and petrochemical operations (9.2%).
Kuala Lumpur, Malaysia	The top five sources of PM_{10} are two-stroke engines (13.1%), industry (17.6%), motor vehicles (42.4%), road dust (18.3%) and soil (8.5%).
Raipur, India	Use of automobiles, the burning of fossil fuels, and industrial processes were the anthropogenic sources of PM.
Kolkata, India	Resuspension of road dust, soil dust, autos, traffic, and surrounding industrial pollutants in the industrial area, as well as coal burning, building activities, and emissions from a solid waste disposal on the residential site, all led to higher PM_{10} pollution.
Navi Mumbai, India	Major sources of coarse particles at an urban location are soil, sea salt, and combustion.
Agra, India	At urban sites, solid waste incineration, industrial emissions, and road dust from vehicle activity were the main sources of PM_{10}, whereas at rural sites, soil dust from vehicular emission, construction activity, and windblown dust were the most frequent sources.
Ahmadabad, India	Mineral dust is a significant source of PM_{10} (43%)

(continued)

Table 1: (continued)

City, Country	*Major sources of PM fraction*
Hyderabad, India	The main causes of PM_{10} pollution at traffic sites were resuspended dust (40%) combustion (12%) and vehicle pollution (22%).
Al Taweela, United Arab Emirates	Major sources of PM_{10} are continental dust (35%), road dust (24%), marine dust (11%), and sea salt (12%).
Wuhan City, China	Construction, demolition, and artificial soil resuspension all contribute 34%, 20%, 16%, 11%, 10%, and 4% of the total PM_{10} emissions from sources related to metallurgy, steel industry and regional anthropogenic background.
Pingdingshan, China	PM_{10} in industrial cities was found to be mostly caused by the weathering of garbage dumps (9.8%), crystalline matter (19%), coal combustion (31.6%), vehicle exhaust and abrasion (7.4%), local burning (6.3%), and industrial metal smelting (25.9%).
Ulaanbaatar, Mongolia	Road dust (12%), vehicle exhaust (3.1%), kiosks (1.9%), brick kilns (1.6%), power plants (24%), heat-only boilers (19%), domestic cook stoves (39%) and open rubbish burning (0.4%) all contribute to PM_{10}.
Sydney, Australia	Major contributors to PM_{10} include crustal matter (24%) sea salt (16%) organics (9.1%) and soot (3.2%).
Brisbane, Australia	A considerable part of the crustal fraction and sea salt in PM_{10} in the coastal area were found to be mostly (37–64%) contributed by vehicular dust.
Cordoba, Argentina	Urban dust and traffic were also determined to be significant sources of PM_{10} mass.
Nelson, New Zealand	Main component of PM_{10} in the suburbs is sea salt (49%) followed by old sea salt (17%), wind-blown soil (16%), nitrate (12%), and road dust (6%).
Arizona, USA	Secondary (3%), organic matter (12%), crustal (54%) and unidentified (14%).
Tijuana, Mexico	Traffic (5%), sea salt (25%), biomass burning (3%), mineral (46%) and unidentified (22%).
Kathmandu Valley, Nepal	Motor vehicle (34%), biomass burning (23%), secondary aerosol (2%) and bricks kiln (6%).
Antofagasta, Chile	Cement plant (38%), marine aerosol (5%), minerals stockpiles (9%), copper smelter (11%) and soil dust (3%).
New Taipei, Taiwan	Vehicle emission (25%), sea salt (8.4%), secondary aerosol (24%), soil dust (34%) and industry (17.6%).

3 Different Methods for Measuring of Oxidative Potential (OP)

3.1 Acellular Method

3.1.1 DTT Assay

Following Kumagai et al. (2002), Cho et al. (2005) described the redox-active chemicals that catalyse the reduction of oxygen to superoxide by DTT (Table 2), which is then oxidised to its disulfide. In DDT assay, pH and temperature (37 °C) are essential factors for maintaining redox condition and incubation of PM extracted samples after the addition of DDT (containing phosphate buffer pH 7.4), respectively. After adding 10% trichloro acetic acid into the incubation mixture and take aliquot. After stopping the reaction with Tris-HCl (pH 8.9) (including 20 mM EDTA), add 5,5′-dithiobis-2-nitrobenzoic acid (DTNB) into the aliquot. The remaining DTT reacts with (DTNB) and generates 2-nitro-5-thiobenzoic acid (TNB). The concentration of TNB was measured by its optical density absorption at 412 nm. According to Gao et al. (2017), DTT substitutes for cellular reductant molecules like NADH/NADPH. Pursuing a purely chemical study, the objective was to imitate interactions between physiological reductants and aerosol constituents. Based on several studies (Cho et al., 2005; Lin and Yu, 2011; Charrier and Anastasio, 2012), DTT consumption over time is related to the amount of PM redox-active species. The oxidative potentials of DTT function as a biomarker, similar to the expression of cellular heme oxygenase (HO-1) in human patients (Li et al., 2003) and fractional exhaled nitric oxide (FENO) (Delfino et al., 2013; Janssen et al., 2015).

3.1.2 Modified DTT Assay

Charrier and Anastasio (2012) followed Cho et al. (2005) procedure for DTT experiments. In this process, the additional step was treating phosphate buffer solutions with Chelex 100 resin to remove transition metals and other impurities in chemicals.

3.1.3 Test for Ascorbic Acid (AA)

The PM extracts are incubated for 10 min at 37°C in order to determine AA depletion (Table 2). Every 2 minutes for 2 hours after administering ascorbic acid, the absorption at 265 nm is measured. Before each measurement, a 96-well plate is automatically shaken for 3 seconds. The linear regression of the linear section data, which is shown as absorbance against time (Fang et al., 2015; Janssen et al., 2014), is used to calculate the maximum depletion rate of AA.

Table 2: Oxidative potential response by different assay

Assay	*Oxidative potential response species*	*References*
DTT	Water soluble and methanol extract PM, OC, EC, PAHs and quinones, metals Fe, Cu and Mn, Zn, CaAl, Co, V and ions K^+	Moufarrej et al., 2020; Calas et al., 2018; Gao et al., 2017; Yang et al., 2014; Carrier and Anastasio, 2012; Ayres et al., 2008; Cho et al., 2005; Fang et al., 2015, 2016; Perrone et al., 2019; Visentin et al., 2016.
AA	Water-soluble metals (especially Cu), traffic-related PM components EC, Fe, Cu and PAHs, Pb, Sr, 1,2-NPQ, 1,4-NPQ and 9,10-(PQN)	Fang et al., 2015, 2016; Janssen et al., 2014; Pietrogrande et al., 2019; Perrone et al., 2019; Visentin et al., 2016; Godri et al., 2010; Calas et al., 2018; Moufarrej et al., 2020.
ESR	Sulfate, nitrate, Fe, Cu, Ni and V, quinones and iron	Janssen et al., 2014; Boogaard et al., 2012, Calas et al., 2018; Arangio et al., 2016; Yang et al., 2014.
GSH	Transitional metal Al, Pb, water-soluble Cu and BrC and fly ash containing nano material TiO_2	Godri et al., 2010; Vejerano et al., 2015; Bates et al., 2019; Gao et al., 2020.

NPQ: 1,2-naphthoquinone, NPQ: 1,4-naphthoquinone, PQN: 9,10-phenantrenequinone, EC and OC: Elemental and organic carbon, BrC: Brown carbon

3.1.4 ESR Assay

According to Shi et al. (2003), Janssen et al. (2014), Yang et al. (2014), and others, the 5,5-dimethyl-1-pyrroline-N-oxide (DMPO) spin trap and cellular Fenton reaction are used to measure the oxidative potential (Table 2) by PM producing hydroxyl radicals (OH) in the presence of hydrogen peroxides (H_2O_2). In simple terms, H_2O_2 and DMPO are blended with PM suspensions, and then the mixture is incubated at 37°C in a heated, agitating water bath. The suspension is vortexed without filtering after incubation and put into a glass capillary. Mini-Scope MS-400 spectrometer measurements of the DMPO-OH quartet signal are made. The average of the total amplitudes of the DMPO-OH quartet in arbitrary units (AU) per g PM is used to compute the electronic susceptibility rate (ESR).

3.1.5 Cellular Assay

a) **DCF assay:** In 1980s there were not enough techniques to measure the ROS. Recent studies have developed many protocols, techniques or methods to quantify the ROS activity in cellular and acellular systems like DCF assay. It is possible to carry out the total ROS quantification measurement (Venkatachari et al., 2005) using a fluorogenic probe like dichlorofluorescin (DCFH) which is a non-fluorescent reagent. In the presence of reactive oxygen species (ROS), DCFH is oxidized to fluorescent dichlorofluorescin (DCF) (Wang and Joseph,

1999; Venkatachari et al., 2005; Møller et al., 2010). The DCF assay, also called fluorometric assay, has been proved to have a high sensitivity and detect picomole levels of hydroperoxides (Wang and Joseph, 1999; Venkatachari et al., 2005). It has been widely used in biological systems as an indicator of oxidative stress in toxicological phenomena. After that, this assay was modified to some extent and applied diverse fields to estimate the generation of reactive oxygen species (Brömme et al., 2008). Recently 2′,7′-dichlorofluorescin ($DCFH_2$) has been used to assay ambient particles generating ROS (Venkatachari et al., 2005; Brömme et al., 2008). The principle is that DCFH reacts with reactive oxygen species (ROS) and reactive nitrogen species (RNS) to generate a fluorescent product (Tuet et al., 2016). In contrast to acellular tests, which call for chemical deacetylation, cellular assays load cells with 2′,7′-dichlorofluoresc in diacetates (DCFH-DA), which is hydrolyzed intracellularly by endogenous esterase. While singlet oxygen, hydrogen peroxide and superoxide anions have a limited ability to oxidise DCFH, the DCFH probe is considered as a measurement of hydroxyl radicals, peroxy nitrite, nitric oxide, and more (Møller et al., 2010; Tuet et al., 2016, 2019).

b) **MTT assay:** The MTT test relies on the formation of formazan crystals by the interaction of mitochondrial dehydrogenase enzymes with the tetrazolium rings of MTT (3-(4,5 dimethylthiazol-2-yl)-2,5-diphenyl tetrazolium bromide) (Tuet et al., 2016; Ng et al., 2019). After being treated with DMSO, healthy cells expel the crystals that have accumulated inside them. A549 cells in the log phase of growth and cell viability inhibition, predictive of cytotoxic effects capable of reducing the mitochondrial activity of cells, were estimated in relative terms taking into account the net impact of PM_{10} concerning blanks to perform the MTT experiment (Lionetto et al., 2021). According to Jeon et al. (2011), the MTT [3-(4,5-dimethylthiazol-2-yl)-2,5-diphenyl tetrazolium bromide] test was used to assess the cytotoxic effects of PM_{10} on human lung epithelial cells. When three different studies were carried out, the PM_{10} extract decreased the cell viability in a dose-dependent manner since exposure to 100 mg/ml of PM10 for 24 h caused around 30% of the cells to die. Similarly, Ng et al. (2019), measured cellular metabolic activity using MTT assay for each ROS/RNS of PM samples. After the incubation period, the tetrazolium dye was reduced by cellular NADPH-dependent oxidoreductases to form formazan (insoluble purple salt). Finally, dimethyl sulfoxide was then added to solubilize the salt and absorbance was measured at 570 nm using a microplate reader (Tuet et al., 2016; Lionetto et al., 2021).

c) **Other assays for oxidative stress determination:** Numerous cellular and acellular assays, such as heme-oxygenase 1 (HO-1) immunoblotting, plasmid scission, GSH/GSSG and cellular stimulation, have been reported throughout the literature (Li et al., 2000; 2002a; 2002b; Li et al., 2003; Vejerano et al., 2015; Shahpoury et al., 2019; Riva et al., 2011 commonly used to evaluate DNA oxidation or damage was the comet test (Lionetto et al., 2021), commonly known as single cell gel electrophoresis. through this experiment, broken DNA fragments travel through the medium while being affected by an electric field, but healthy

DNA migrates far more slowly. By analysing the DNA "comet" tail form and migratory pattern, the degree of DNA damage is determined.

3.2 PM Induced Oxidative Stress

The PM is now a concerning matter globally due to its adverse health effects. It disrupts the balance between reactive oxygen species (ROS) or reactive nitrogen species (RNS) and antioxidant defences, called oxidative stress (Fig. 1). The imbalance of free radicals (ROS) and antioxidants in the body can damage cells and tissues. In producing high concentrations of ROS in cells, the defence mechanism completely fail and is considered important mechanism of particle-induced health effects. The particles containing free radicals (ROS) acts as a scavenger in the body system; as a result, to start the oxidation of cell materials like lipid and DNA. The lipid peroxidation products and oxidatively produced DNA lesions are often used as biomarkers of oxidative stress in humans and animals (Moller et al., 2010). For the human study to investigate oxidative stress, to mainly use surrogate tissue such as leukocytes, plasma and urine for OP assessment. The nano sized particles or ultrafine particles can translocate across the air–blood barrier because of their distribution pattern in the lung, resistance to phagocytosis and small size enabling passage (Donaldson et al., 2001).

Numerous epidemiological studies have found the association between exposure to particles and oxidative stress shows cardiovascular effects and cancer. According to Nel et al. (2006), particle-make oxidative stress affects cell signalling in three tiers with enhanced transcription of defence genes through the transcription factor Nrf2 (nuclear factor erythroid 2–related factor 2; protect against oxidative damage) at low levels of oxidative stress. At the highest level of oxidative stress occurs the activation of inflammation signalling through NFκB (nuclear factor kappa B) at higher levels and activation of apoptotic pathways and necrosis of the cell system. Oxidative stress (OS) has been a concern in various diseases in ageing such as atherosclerosis, cancer and neurodegenerative diseases such as Parkinson's disease, Alzheimer's, etc. (Wang and Joseph., 2006). In addition, a recent study found that the positive association of $PM_{2.5}$ and neurodegenerative diseases such as Alzheimer's disease (AD), non-AD dementia, and Parkinson's disease (PD) are increasingly common among older adults (Rhew et al., 2021; Jankowska-Kieltyka et al., 2021). *In vitro* and *in vivo* studies have proposed that the fraction of the coarse particles shows toxic effects that are at least as potent and sometimes more (mass basis) than fine PM (Kelly and Fussell, 2012). It shows greater toxicity, oxidant production and cytokine release from human monocytes and alveolar macrophages (Monn and Becker, 1999; Soukup and Becker, 2001).

On the other hand, Chuang et al. (2005) found the cardiovascular effect of smaller particles with an association between reduced heart rate variability (HRV) and $PM_{0.3\text{-}1.0}$. The exposure to $PM_{2.5}$, the sympathetic nervous system is activated, and the hypothalamus becomes inflamed, which amplifies the blood pressure response

(Munzel et al., 2018). According to Steenhof et al. (2011), the proinflammatory (tumor necrosis factor-alpha, TNF-α; Interleukin-6, IL-6; Macrophage Inflammatory Protein-2, MIP-2) response was highest for fine PM than coarse PM and caused a concentration-dependent decrease in MTT-reduction activity and increase in production of TNF-α, IL-6 and MIP-2. Becker et al. (2005) also found that cytokine induction by coarse PM induced a greater increase in IL-6 release compared with fine and ultrafine PM in alveolar macrophages (AMs) and normal human bronchial epithelial (NHBE) cells. The particles bound benzo(a) pyrene (BaP) can induce DNA oxidation and damage in systemic target organs, including lung and kidney (Gerde et al., 2001; Kim and Lee, 1997). In addition, Li et al. (2003) proposed that the ultrafine PM penetrates the tissue and generates oxidative stress at more damaging cellular level and consequently contributes to the adverse health effects. A study found that PM_{10} decreased the level of ezrin (a membrane-actin cytoskeleton linker protein) expression, it is a multidomain protein providing regulated membrane–cytoskeleton contacts that play a role in the regulation of the growth, adhesion, differentiation and metastatic behavior of cancer cells (Austermann et al., 2008; Park et al., 2009; Wang et al., 2009). The apoptotic (process of programmed cell death) rate for tumor cells within the lung, the metastasis was low in cells with high ezrin, whereas, in cells with suppressed ezrin, the apoptotic rate was significantly higher. TNF receptor-associated protein 1 (TRAP1) is a mitochondrial heat-shock protein known to play a role in protection against oxidative stress and apoptosis, but it was down-regulated in PM_{10}-treated cells (Jeon et al., 2011).

3.3 PM Containing Transition Metals and Oxidative Stress

According to their source apportionment, PM containing transition metals in the atmosphere might vary, for example, copper component discovered in automobile brake pads, ceramic brake pads (10–20%), and metallic brake pads included considerable iron (70%) and very little copper (Lodovici and Bigagli, 2011). Transition metals including iron (Fe), lead (Pb), mercury (Hg), cadmium (Cd), silver (Ag), nickel (Ni), vanadium (V), chromium (Cr), manganese (Mn), copper (Cu) and zinc (Zn) are capable of producing ROS (Gerlofs-Nijland et al., 2007; Kim et al., 2021; Sørensen et al., 2005). Kam et al. (2011) found that reactive oxygen species (ROS) are strongly correlated with water-soluble Fe, Ni and organic carbon (OC). The water-soluble fraction of PM such as Fe, Ni, V, Co, Cu, Cr etc. causes DNA damage via transition metal-dependent ʻOH formation by Fenton reaction (Fig. 3) where H_2O_2 plays an important role (Knaapen et al., 2002; Jomova and Valko, 2011). A short communication by Sørensen et al. (2005), found the soluble transition metals ($PM_{2.5}$) such as V, Cr, Fe, Ni, Cu and platinum associated with DNA damage in urine samples in terms of 7-hydro-8-oxo-2V-deoxyguanosine (8-oxodG). They also found that the 8-oxod G concentration in lymphocytes was significantly associated with the V and Cr concentrations with a 1.9% and 2.2% increase in 8-oxodG per 1 Mg/L increase in the V and Cr concentration, respectively. The major source of V is the combustion of

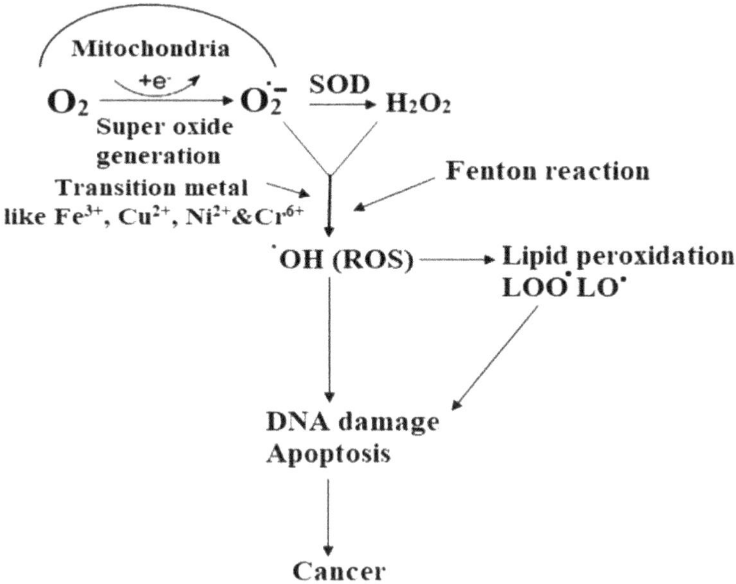

Figure 3: Possible mechanisms of PM containing transitional metal induced ROS generation (by Fenton reaction) and cell apoptosis.

heavy oil and Cr is originating mainly from the wear of brake linings. Becker et al. (2005) found the association of individual PM elements with cytokine release. The PM elements such as Fe and Si were positively associated with IL-6 release in AMs, whereas Cr was positively associated with IL-8 release in NHBE cells.

4 PM and Health Effects

The PM is significantly associated with various adverse health effects (Table 3) such as childhood respiratory illness (Yang et al., 2004), chronic obstructive pulmonary disease (COPD) and chronic bronchitis (Kim and Criner, 2013), ischemic heart disease and oxidative stress (Wang and Joseph., 2006; Li et al., 2003; Araujo et al., 2008; Kim et al., 2021; Sørensen et al., 2005), atherosclerosis (Brook and Rajagopalan, 2010), lipid peroxidation (Pereira et al., 2007), mitochondrial damage (Li et al., 2003), vascular inflammation (Gong et al., 2007), myocardial infarction (MI; Weichenthal et al., 2016), blood pressure and incident hypertension (Zhang et al., 2018, 2019; Li et al., 2019) and also DNA oxidation (Sørensen et al., 2005). The various study found that exposure to PM_1, $PM_{2.5}$ and PM_{10} was associated with increased blood pressure and hypertension in children and adolescents of China (Zhang et al., 2018; Zhang et al., 2019; Li et al., 2019); where PM_{10} exposer was also associated with a higher prevalence of hypertension (Zhang et al., 2019) and long-term exposure to ambient PM_1 increases the risk of hypertension and with elevations of blood pressure (Li et al., 2019). The organic fraction of PM containing benzo[a]pyrene (PAHs) can induce DNA oxidation and damage in systemic target organs and it is classified as a class-1 carcinogen for their carcinogenicity by the International Agency for Research on Cancer (IARC). The $PM_{2.5}$ is associated with increase cardiovascular morbidity and mortality (Brook and Rajagopalan, 2010). In addition, short-term $PM_{2.5}$ exposure triggers acute cardiovascular events and longer-term exposure increases the cardiovascular risk to a greater extent and may promote the progression of atherosclerotic plaques (Araujo, 2011). Weichenthal et al. (2016) found the strongest associations between ambient $PM_{2.5}$ and emergency hospital visits for myocardial infarction. In China, the more significant short-term effects show PM_1 in comparison to $PM_{2.5}$ and PM_{10} on childhood pneumonia (Wang et al., 2021). Wu et al. (2021) reported that short-term exposure to $PM_{2.5}$ (daily T2DM deaths; 1.099%), PM_{10} (1.016%), SO_2 (3.835%), and NO_2 (1.587%) were positively associated with Type 2 diabetes mellitus (T2DM) mortality where gaseous pollutants have greater effects than particulate matter (PM).

5 Conclusions

The ambient PM and containing metals are great threat on human health due to their numerous effects on organs or cell. The PM and/or containing metals are potent to various short- and long-term diseases such as blood pressure, eye irritation, sneezing, coughing, pneumonia, hyperacidity, sinusitis, bronchitis, asthma, COPD, lung inflammation and injury, reduction of lung function capacity, cardiovascular diseases and lung cancer. The various epidemiological and clinical studies prove the association of PM and various health effects. More research and advanced techniques and methods are needed to find the actual mechanisms or pathways about how PM and its containing metals generates ROS into the cell.

Table 3: Possible health effects caused by aggregate PM and containing metals

PM	*Health effects*	*References*
Coarse, fine and ultrafine PM	• Ambient particles affect eye irritation, respiratory disorder (nose block, sneezing, cough, hyperacidity, sinusitis, bronchiti and asthma), lung inflammation and injury, reduced lung function capacity, pneumonia, infant death syndrome, cardiovascular diseases. • PM_{10} is associated with ischemic heart diseases and myocardial infarction among the elderly population. • Fine particles may cause alveolar inflammation, reduced heart rate, mild hypoxemia or decline in blood oxygen saturation, increased symptoms of obstructive airway disease (such as COPD, bronchitis, and chest illness). High concentration of $PM_{2.5}$ exposure is strongly associated with myocardial infarction (MI), ischemic heart disease, dysrhythmias, heart failure, cardiac arrest and T2DM mortality. • Ultrafine particles can penetrate blood barriers and affect the vascular system as inflammation, oxidative stress, cellular damage, mitochondrial damage, lipid peroxidation, and DNA oxidation.	Yang et al., 2004; Kim and Criner, 2013; Wang and Joseph, 2006; Li et al., 2003; Araujo et al., 2008; Rai, 2015; Kim et al., 2021; Sørensen et al., 2005; Li et al., 2003; Araujo, 2011; Weichenthal et al., 2016; Wang et al., 2021; Wu et al., 2021; Jeon et al., 2011; Gerde et al., 2001; Kim and Lee, 1997; Becker et al., 2005; Knaapen et al., 2002; Gerlofs-Nijland et al., 2007; Kim et al., 2021; Sørensen et al., 2005.
Transition metal	Transition metals such as iron, lead, mercury, cadmium, silver, nickel, vanadium, chromium, manganese, and copper, Zinc are capable of ROS formation causes of protein denaturation resulting of cell damage or apoptosis. Fe and Si were positively associated with IL-6 and IL-8 release in the cell. It also causes many health effects such as cancer, neurotoxicity, immune toxicity and cardiotoxicity.	

Acknowledgement B.Ghosh would like to thank the University Grants Commission, New Delhi, for providing Non-NET fellowship during this study.

References

Adams K, Greenbaum DS, Shaikh R, van Erp AM, Russell AG (2015) Particulate matter components, sources, and health: systematic approaches to testing effects. J Air Waste Manag Assoc 65(5):544–558

Adler V, Yin Z, Tew KD, Ronai ZE (1999) Role of redox potential and reactive oxygen species in stress signaling. Oncogene 18(45):6104–6111

ALA (2020) American lung association. https://www.lung.org/clean-air/outdoors/what-makes-air-unhealthy/particle-pollution. Accessed 20 July 2021

Arangio AM, Tong H, Socorro J, Pöschl U, Shiraiwa M (2016) Quantification of environmentally persistent free radicals and reactive oxygen species in atmospheric aerosol particles. Atmos Chem Phys 16(20):13105–13119. https://doi.org/10.5194/acp-16-13105-2016

Araujo JA (2011) Particulate air pollution, systemic oxidative stress, inflammation, and atherosclerosis. Air Qual Atmos Health 4(1):79–93

Araujo JA, Barajas B, Kleinman M, Wang X, Bennett BJ, Gong KW, Navab M, Harkema J, Sioutas C, Lusis AJ, Nel AE (2008) Ambient particulate pollutants in the ultrafine range promote early atherosclerosis and systemic oxidative stress. Circ Res 102(5):589–596

Austermann J, Nazmi AR, Müller-Tidow C, Gerke V (2008) Characterization of the Ca^{2+}-regulated Ezrin-S100P interaction and its role in tumor cell migration*. J Biol Chem 283(43):29331–29340

Ayres JG, Borm P, Cassee FR, Castranova V, Donaldson K, Ghio A, Maynard RL, Mudway I, Nel A, Sioutas C, Smith S, Baeza-Squiban A, Cho A, Duggan S, Froines J (2008) Evaluating the toxicity of airborne particulate matter and nanoparticles by measuring oxidative stress potential—a workshop report and consensus statement. Inhal Toxicol 20(1):75–99

Bates JT, Fang T, Verma V, Zeng L, Weber RJ, Tolbert PE, Abrams JY, Sarnat SE, Klein M, Mulholland JA, Russell AG (2019) Review of acellular assays of ambient particulate matter oxidative potential: methods and relationships with composition, sources, and health effects. Environ Sci Technol 53(8):4003–4019

Baulig A, Poirault JJ, Ausset P, Schins R, Shi T, Baralle D, Dorlhene P, Meyer M, Lefevre R, Baeza-Squiban A, Marano F (2004) Physicochemical characteristics and biological activities of seasonal atmospheric particulate matter sampling in two locations of Paris. Environ Sci Technol 38(22):5985–5992

Becker S, Dailey LA, Soukup JM, Grambow SC, Devlin RB, Huang YCT (2005) Seasonal variations in air pollution particle-induced inflammatory mediator release and oxidative stress. Environ Health Perspect 113(8):1032–1038

Bondareva AA, Capecchi MR, Iverson SV, Li Y, Lopez NI, Lucas O, Merrill GF, Prigge JR, Siders AM, Wakamiya M, Wallin SL, Schmidt EE (2007) Effects of thioredoxin reductase-1 deletion on embryogenesis and transcriptome. Free Radic Biol Med 43(6):911–923

Boogaard H, Janssen NA, Fischer PH, Kos GP, Weijers EP, Cassee FR, van der Zee SC, de Hartog JJ, Brunekreef B, Hoek G (2012) Contrasts in oxidative potential and other particulate matter characteristics collected near major streets and background locations. Environ Health Perspect 120(2):185–191

Brömme HJ, Zühlke L, Silber RE, Simm A (2008) DCFH2 interactions with hydroxyl radicals and other oxidants–influence of organic solvents. Exp Gerontol 43(7):638–644

Brook RD, Rajagopalan S (2010) Particulate matter air pollution and atherosclerosis. Curr Atheroscler Rep 12(5):291–300

Calas A, Uzu G, Kelly FJ, Houdier S, Martins JM, Thomas F, Molton F, Charron A, Dunster C, Oliete A, Jacob V, Besombes J-L, Chevrier V, Jaffrezo JL (2018) Comparison between five acellular oxidative potential measurement assays performed with detailed chemistry on PM_{10} samples from the city of Chamonix (France). Atmos Chem Phys 18(11):7863–7875

Castelo FB (2017) Human exposure assessment related to oil activities in Ecuador: From the air quality monitoring to the study of metallic contaminants transfer in the soil-plant continuum. Doctoral dissertation. Universite Toulouse 3 Paul Sabatier (UT3 Paul Sabatier)

CDC (2019) Centers for disease control and prevention. https://www.cdc.gov/air/particulate_matter.html. Accessed 22 July 2021

Charrier JG, Anastasio C (2012) On dithiothreitol (DTT) as a measure of oxidative potential for ambient particles: evidence for the importance of soluble transition metals. Atmos Chem Phys (Print) 12(5):11317

Cho AK, Sioutas C, Miguel AH, Kumagai Y, Schmitz DA, Singh M, Eiguren-Fernandez A, Froines JR (2005) Redox activity of airborne particulate matter at different sites in the Los Angeles Basin. Environ Res 99(1):40–47

Chuang K-J, Chan C-C, Chen N-T, Su T-C, Lin L-Y (2005) Effects of particle size fractions on reducing heart rate variability in cardiac and hypertensive patients. Environ Health Perspect 113:1693–1697

Crobeddu B, Aragao-Santiago L, Bui LC, Boland S, Squiban AB (2017) Oxidative potential of particulate matter 2.5 as predictive indicator of cellular stress. Environ Pollut 230:125–133

Dagher Z, Garçon G, Billet S, Verdin A, Ledoux F, Courcot D, Aboukais A, Shirali P (2007) Role of nuclear factor-kappa B activation in the adverse effects induced by air pollution particulate matter ($PM_{2.5}$) in human epithelial lung cells (L132) in culture. J Appl Toxicol: Int J 27(3):284–290

Delfino RJ, Staimer N, Vaziri ND (2011) Air pollution and circulating biomarkers of oxidative stress. Air Qual Atmos Health 4(1):37–52

Delfino RJ, Staimer N, Tjoa T, Gillen DL, Schauer JJ, Shafer MM (2013) Airway inflammation and oxidative potential of air pollutant particles in a pediatric asthma panel. J Expo Sci Environ Epidemiol 23(5):466–473

Donaldson K, Stone V, Clouter A, Renwick L, MacNee W (2001) Ultrafine particles. Occup Environ Med 58(3):211–216

Fang T, Fu LZ, Zhou LL, Zhan SZ (2015) A water-soluble dinuclear copper electrocatalyst, [Cu (oxpn) Cu $(OH)_2$] for both water reduction and oxidation. Electrochim Acta 161:388–394

Fang T, Guo H, Zeng L, Verma V, Nenes A, Weber RJ (2017) Highly acidic ambient particles, soluble metals, and oxidative potential: a link between sulfate and aerosol toxicity. Environ Sci Technol 51(5):2611–2620

Fang T, Verma V, Bates JT, Abrams J, Klein M, Strickland MJ, Sarnat SE, Chang HH, Mulholland JA, Tolbert PE, Russell AG, Weber RJ (2016) Oxidative potential of ambient water-soluble PM_{25} in the southeastern United States: contrasts in sources and health associations between ascorbic acid (AA) and dithiothreitol (DTT) assays. Atmos Chem Phys 16(6):3865–3879

Gao D, Fang T, Verma V, Zeng L, Weber RJ (2017) A method for measuring total aerosol oxidative potential (OP) with the dithiothreitol (DTT) assay and comparisons between an urban and roadside site of water-soluble and total OP. Atmos Meas Techn 10(8):2821–2835

Gao D, Godri Pollitt KJ, Mulholland JA, Russell AG, Weber RJ (2020) Characterization and comparison of $PM_{2.5}$ oxidative potential assessed by two acellular assays. Atmos Chem Phys 20(9):5197–5210

Gerde P, Muggenburg BA, Lundborg M, Tesfaigzi Y, Dahl AR (2001) Respiratory epithelial penetration and clearance of particle-borne benzo [a] pyrene. Research Report-Health Effects Institute

Gerlofs-Nijland ME, Dormans JA, Bloemen HJ, Leseman DL, Boere AJ, Kelly FJ, Mudway IS, Jimenez AA, Donaldson K, Guastadisegni C, Janssen NA, Brunekreef B, Sandström T, van Bree L, Cassee FR (2007) Toxicity of coarse and fine particulate matter from sites with contrasting traffic profiles. Inhal Toxicol 19(13):1055–1069. https://doi.org/10.1080/08958370701626261

Godri KJ, Duggan ST, Fuller GW, Baker T, Green D, Kelly FJ, Mudway IS (2010) Particulate matter oxidative potential from waste transfer station activity. Environ Health Perspect 118(4):493–498

Gong KW, Zhao W, Li N, Barajas B, Kleinman M, Sioutas C, Horvath S, Lusis AJ, Nel A, Araujo JA (2007) Air-pollutant chemicals and oxidized lipids exhibit genome-wide synergistic effects on endothelial cells. Genome Biol 8(7):1–13

González-Flecha B (2004) Oxidant mechanisms in response to ambient air particles. Mol Aspects Med 25(1–2):169–182

Gordon T, Lippmann M, Nádas A, Hickey C (2013) NPACT study 2: *Vitro* and *In vivo* toxicity of exposure to coarse, fine, and ultrafine PM from five airsheds. In: National particle component toxicity (NPACT) initiative: integrated epidemiologic and toxicologic studies of the health effects of particulate matter components, pp 55-94.

Hamanaka RB, Mutlu GM (2018) Particulate matter air pollution: effects on the cardiovascular system. Front Endocrinol 9:680

Harrison RM, Deacon AR, Jones MR, Appleby RS (1997) Sources and processes affecting concentrations of PM_{10} and $PM_{2.5}$ particulate matter in Birmingham (UK). Atmos Environ 31(24):4103–4117

Hatfield DL, Yoo MH, Carlson BA, Gladyshev VN (2009) Selenoproteins that function in cancer prevention and promotion. Biochim Biophys Acta (BBA) – Gen Subj 1790(11):1541–1545

Hetland RB, Cassee FR, Låg M, Refsnes M, Dybing E, Schwarze PE (2005) Cytokine release from alveolar macrophages exposed to ambient particulate matter: heterogeneity in relation to size, city and season. Part Fibre Toxicol 2(1):1–15

Hetland RB, Cassee FR, Refsnes M, Schwarze PE, Låg M, Boere AJF, Dybing E (2004) Release of inflammatory cytokines, cell toxicity and apoptosis in epithelial lung cells after exposure to ambient air particles of different size fractions. Toxicol in Vitro 18(2):203–212

Jankowska-Kieltyka M, Roman A, Nalepa I (2021) The air we breathe: air pollution as a prevalent proinflammatory stimulus contributing to neurodegeneration. Front Cell Neurosci 15:239

Janssen NA, Strak M, Yang A, Hellack B, Kelly FJ, Kuhlbusch TA, Harrison RM, Brunekreef B, Cassee FR, Steenhof M, Hoek G (2015) Associations between three specific a-cellular measures of the oxidative potential of particulate matter and markers of acute airway and nasal inflammation in healthy volunteers. Occup Environ Med 72(1):49–56

Janssen NA, Yang A, Strak M, Steenhof M, Hellack B, Gerlofs-Nijland ME, Kuhlbusch T, Kelly F, Harrison R, Brunekreef B, Hoek G, Cassee F (2014) Oxidative potential of particulate matter collected at sites with different source characteristics. Sci Total Environ 472:572–581

Jeon YM, Son BS, Lee MY (2011) Proteomic identification of the differentially expressed proteins in human lung epithelial cells by airborne particulate matter. J Appl Toxicol 31(1):45–52

Johnson RL Jr (2004) Relative effects of air pollution on lungs and heart. Circulation 109(1):5–7

Jomova K, Valko M (2011) Advances in metal-induced oxidative stress and human disease. Toxicology 283(2–3):65–87

Jung H, Guo B, Anastasio C, Kennedy IM (2006) Quantitative measurements of the generation of hydroxyl radicals by soot particles in a surrogate lung fluid. Atmos Environ 40(6):1043–1052

Kam W, Ning Z, Shafer MM, Schauer JJ, Sioutas C (2011) Chemical characterization and redox potential of coarse and fine particulate matter (PM) in underground and ground-level rail systems of the Los Angeles Metro. Environ Sci Technol 45(16):6769–6776

Kelly FJ, Fussell JC (2012) Size, source and chemical composition as determinants of toxicity attributable to ambient particulate matter. Atmos Environ 60:504–526

Kim HJ, Herath KHINM, Dinh DTT, Kim HS, Jeon YJ, Kim HJ, Jee Y (2021) Sargassum horneri ethanol extract containing polyphenols attenuates PM-induced oxidative stress via ROS scavenging and transition metal chelation. J Funct Foods 79:104401

Kim KB, Lee BM (1997) Oxidative stress to DNA, protein, and antioxidant enzymes (superoxide dismutase and catalase) in rats treated with benzo (a) pyrene. Cancer Lett 113(1–2):205–212

Kim S, Jaques PA, Chang M, Froines JR, Sioutas C (2001) Versatile aerosol concentration enrichment system (VACES) for simultaneous in vivo and in vitro evaluation of toxic effects of ultrafine, fine and coarse ambient particles part I: development and laboratory characterization. J Aerosol Sci 32(11):1281–1297

Kim V, Criner GJ (2013) Chronic bronchitis and chronic obstructive pulmonary disease. Am J Respir Crit Care Med 187(3):228–237

Knaapen AM, Shi T, Borm PJ, Schins RP (2002) Soluble metals as well as the insoluble particle fraction are involved in cellular DNA damage induced by particulate matter. In: oxygen/nitrogen radicals: cell injury and disease. Springer, Boston, MA, pp 317–326. https://doi.org/10.1007/978-1-4615-1087-1_36

Kumagai Y, Koide S, Taguchi K, Endo A, Nakai Y, Yoshikawa T, Shimojo N (2002) Oxidation of proximal protein sulfhydryls by phenanthraquinone, a component of diesel exhaust particles. Chem Res Toxicol 15(4):483–489

Li N, Chen G, Liu F, Mao S, Liu Y, Hou Y, Lu Y, Liu S, Wang C, Xiang H, Guo Y, Li S (2019) Associations of long-term exposure to ambient PM1 with hypertension and blood pressure in rural Chinese population: the Henan rural cohort study. Environ Int 128:95–102

Li N, Kim S, Wang M, Froines J, Sioutas C, Nel A (2002a) Use of a stratified oxidative stress model to study the biological effects of ambient concentrated and diesel exhaust particulate matter. Inhal Toxicol 14(5):459–486

Li N, Sioutas C, Cho A, Schmitz D, Misra C, Sempf J, Wang M, Oberley T, Froines J, Nel A (2003) Ultrafine particulate pollutants induce oxidative stress and mitochondrial damage. Environ Health Perspect 111(4):455–460

Li N, Venkatesan MI, Miguel A, Kaplan R, Gujuluva C, Alam J, Nel A (2000) Induction of heme oxygenase-1 expression in macrophages by diesel exhaust particle chemicals and quinones via the antioxidant-responsive element. J Immunol 165(6):3393–3401

Li N, Wang M, Oberley TD, Sempf JM, Nel AE (2002b) Comparison of the pro-oxidative and proinflammatory effects of organic diesel exhaust particle chemicals in bronchial epithelial cells and macrophages. J Immunol 169(8):4531–4541

Li XY, Hao L, Liu YH, Chen CY, Pai VJ, Kang JX (2017) Protection against fine particle-induced pulmonary and systemic inflammation by omega-3 polyunsaturated fatty acids. Biochim Biophys Acta (BBA) – Gen Subj 1861(3):577–584

Lin P, Yu JZ (2011) Generation of reactive oxygen species mediated by humic-like substances in atmospheric aerosols. Environ Sci Technol 45(24):10362–10368. https://doi.org/10.1021/es2028229

Lionetto MG, Guascito MR, Giordano ME, Caricato R, De Bartolomeo AR, Romano MP, Conte M, Dinoi A, Contini D (2021) Oxidative potential, cytotoxicity, and intracellular oxidative stress generating capacity of PM_{10}: a case study in South of Italy. Atmosphere 12(4):464

Lodovici M, Bigagli E (2011) Oxidative stress and air pollution exposure. J Toxicol

Møller P, Jacobsen NR, Folkmann JK, Danielsen PH, Mikkelsen L, Hemmingsen JG, Vesterdal LK, Forchhammer L, Wallin H, Loft S (2010) Role of oxidative damage in toxicity of particulates. Free Radic Res 44(1):1–46

Monn C, Becker S (1999) Cytotoxicity and induction of proinflammatory cytokines from human monocytes exposed to fine ($PM_{2.5}$) and coarse particles ($PM_{10-2.5}$) in outdoor and indoor air. Toxicol Appl Pharmacol 155(3):245–252

Moufarrej L, Courcot D, Ledoux F (2020) Assessment of the $PM_{2.5}$ oxidative potential in a coastal industrial city in Northern France: relationships with chemical composition, local emissions and long range sources. Sci Total Environ 748:141448. https://doi.org/10.1016/j.scitotenv.2020.1414

Mudway IS, Duggan ST, Venkataraman C, Habib G, Kelly FJ, Grigg J (2005) Combustion of dried animal dung as biofuel results in the generation of highly redox active fine particulates. Part Fibre Toxicol 2(1):1–11

Mukherjee A, Agrawal M (2017) World air particulate matter: sources, distribution and health effects. Environ Chem Lett 15(2):283–309

Münzel T, Gori T, Al-Kindi S, Deanfield J, Lelieveld J, Daiber A, Rajagopalan S (2018) Effects of gaseous and solid constituents of air pollution on endothelial function. Eur Heart J 39(38):3543–3550

Nel AE, Diaz-Sanchez D, Li N (2001) The role of particulate pollutants in pulmonary inflammation and asthma: evidence for the involvement of organic chemicals and oxidative stress. Curr Opin Pulm Med 7(1):20–26

Nel A, Xia T, Mädler L, Li N (2006) Toxic potential of materials at the nanolevel. Science 311(5761):622–627

Ng NL, Tuet WY, Chen Y, Fok S, Gao D, Rodriguez MS, Klein M, Grosberg A, Weber RJ, Champion JA (2019) Cellular and acellular assays for measuring oxidative stress induced by ambient and laboratory-generated aerosols. Research Reports: Health Effects Institute

Osornio-Vargas ÁR, Bonner JC, Alfaro-Moreno E, Martínez L, García-Cuellar C, Ponce-de-León Rosales S, Miranda J, Rosas I (2003) Proinflammatory and cytotoxic effects of Mexico City air pollution particulate matter in vitro are dependent on particle size and composition. Environ Health Perspect 111(10):1289–1293

Osornio-Vargas AR, Serrano J, Rojas-Bracho L, Miranda J, García-Cuellar C, Reyna MA, Flores G, Zuk M, Quintero M, Vázquez I, Sánchez-Pérez Y, López T, Rosas I (2011) In vitro biological effects of airborne $PM_{2.5}$ and PM_{10} from a semi-desert city on the Mexico–US border. Chemosphere 83(4):618–626

Ostro BD, Broadwin R, Lipsett MJ (2000) Coarse and fine particles and daily mortality in the Coachella Valley, California: a follow-up study. J Expo Sci & Environ Epidem 10(5):412–419

Park EJ, Park K (2009) Oxidative stress and pro-inflammatory responses induced by silica nanoparticles in vivo and in vitro. Toxicol Lett 184(1):18–25

Pereira CEL, Heck TG, Saldiva PHN, Rhoden CR (2007) Ambient particulate air pollution from vehicles promotes lipid peroxidation and inflammatory responses in rat lung. Braz J Med Biol Res 40:1353–1359

Perrone MR, Bertoli I, Romano S, Russo M, Rispoli G, Pietrogrande MC (2019) $PM_{2.5}$ and PM_{10} oxidative potential at a Central Mediterranean Site: contrasts between dithiothreitol-and ascorbic acid-measured values in relation with particle size and chemical composition. Atmos Environ 210:143–155. https://doi.org/10.1016/j.atmosenv.2019.04.04

Pietrogrande MC, Bertoli I, Manarini F, Russo M (2019) Ascorbate assay as a measure of oxidative potential for ambient particles: evidence for the importance of cell-free surrogate lung fluid composition. Atmos Environ 211:103–112. https://doi.org/10.1016/j.atmosenv.2019.05.01

Pipalatkar P, Khaparde VV, Gajghate DG, Bawase MA (2014) Source apportionment of $PM_{2.5}$ using a CMB model for a centrally located Indian city. Aerosol Air Qual Res 14(3):1089–1099

Pope CA III, Burnett RT, Thun MJ, Calle EE, Krewski D, Ito K, Thurston GD (2002) Lung cancer, cardiopulmonary mortality, and long-term exposure to fine particulate air pollution. JAMA 287(9):1132–1141

Pope CA III, Burnett RT, Thurston GD, Thun MJ, Calle EE, Krewski D, Godleski JJ (2004) Cardiovascular mortality and long-term exposure to particulate air pollution: epidemiological evidence of general pathophysiological pathways of disease. Circulation 109(1):71–77

Pope CA III, Burnett RT, Krewski D, Jerrett M, Shi Y, Calle EE, Thun MJ (2009) Cardiovascular mortality and exposure to airborne fine particulate matter and cigarette smoke: shape of the exposure-response relationship. Circulation 120(11):941–948

Pope CA, Ezzati M, Cannon JB, Allen RT, Jerrett M, Burnett RT (2018) Mortality risk and $PM_{2.5}$ air pollution in the USA: an analysis of a national prospective cohort. Air Qual Atmos & Health 11(3):245–252

Qin Y, Kim E, Hopke PK (2006) The concentrations and sources of $PM_{2.5}$ in metropolitan New York City. Atmos Environ 40:312–332

Querol X, Alastuey A, Rodriguez S, Plana F, Ruiz CR, Cots N, Massagué G, Puig O (2001) PM_{10} and $PM_{2.5}$ source apportionment in the Barcelona Metropolitan area, Catalonia, Spain. Atmos Environ 35(36):6407–6419

Rai PK (2015) Multifaceted health impacts of particulate matter (PM) and its management: an overview. Environ Skept Crit 4(1):1

Rhew SH, Kravchenko J, Lyerly HK (2021) Exposure to low-dose ambient fine particulate matter $PM_{2.5}$ and Alzheimer's disease, non-Alzheimer's dementia, and Parkinson's disease in North Carolina. PLoS ONE 16(7):e0253253

Riva D, Magalhães CB, Lopes A, Lanças T, Mauad T, Malm O, Valença SS, Saldiva PH, Faffe DS, Zin WA (2011) Low dose of fine particulate matter ($PM_{2.5}$) can induce acute oxidative stress, inflammation and pulmonary impairment in healthy mice. Inhal Toxicol 23(5):257–267

Sánchez-Pérez Y, Chirino YI, Osornio-Vargas ÁR, Morales-Bárcenas R, Gutiérrez-Ruíz C, Vázquez-López I, García-Cuellar CM (2009) DNA damage response of A549 cells treated with particulate matter (PM_{10}) of urban air pollutants. Cancer Lett 278(2):192–200

Schafer FQ, Buettner GR (2001) Redox environment of the cell as viewed through the redox state of the glutathione disulfide/glutathione couple. Free Radic Biol Med 30(11):1191–1212

Shahpoury P, Harner T, Lammel G, Lelieveld S, Tong H, Wilson J (2019) Development of an antioxidant assay to study oxidative potential of airborne particulate matter. Atmos Meas Techn 12(12):6529–6539

Shi T, Schins RP, Knaapen AM, Kuhlbusch T, Pitz M, Heinrich J, Borm PJ (2003) Hydroxyl radical generation by electron paramagnetic resonance as a new method to monitor ambient particulate matter composition. J Environ Monit 5(4):550–556

Sørensen M, Schins RP, Hertel O, Loft S (2005) Transition metals in personal samples of $PM_{2.5}$ and oxidative stress in human volunteers. Cancer Epidemiol Prev Biomark 14(5):1340–1343

Soukup JM, Becker S (2001) Human alveolar macrophage responses to air pollution particulates are associated with insoluble components of coarse material, including particulate endotoxin. Toxicol Appl Pharmacol 171:20–26

Squadrito GL, Cueto R, Dellinger B, Pryor WA (2001) Quinoid redox cycling as a mechanism for sustained free radical generation by inhaled airborne particulate matter. Free Radic Biol Med 31(9):1132–1138

Steenhof M, Gosens I, Strak M, Godri KJ, Hoek G, Cassee FR, Mudway IS, Kelly FJ, Harrison RM, Lebret E, Brunekreef B, Janssen NAH, Pieters RH (2011) In vitro toxicity of particulate matter (PM) collected at different sites in the Netherlands is associated with PM composition, size fraction and oxidative potential – the RAPTES project. Part Fibre Toxicol 8(1):1–15

Tuet WY, Fok S, Verma V, Rodriguez MST, Grosberg A, Champion JA, Ng NL (2016) Dose-dependent intracellular reactive oxygen and nitrogen species (ROS/RNS) production from particulate matter exposure: comparison to oxidative potential and chemical composition. Atmos Environ 144:335–344

Tuet WY, Liu F, de Oliveira AN, Fok S, Artaxo P, Vasconcellos P, Champion JA, Ng NL (2019) Chemical oxidative potential and cellular oxidative stress from open biomass burning aerosol. Environ Sci Technol Lett 6(3):126–132

Turner MC, Krewski D, Pope CA III, Chen Y, Gapstur SM, Thun MJ (2011) Long-term ambient fine particulate matter air pollution and lung cancer in a large cohort of never-smokers. Am J Respir Crit Care Med 184(12):1374–1381

Vejerano EP, Ma Y, Holder AL, Pruden A, Elankumaran S, Marr LC (2015) Toxicity of particulate matter from incineration of nanowaste. Environ Sci Nano 2(2):143–154

Venkatachari P, Hopke PK, Grover BD, Eatough DJ (2005) Measurement of particle-bound reactive oxygen species in Rubidoux aerosols. J Atmos Chem 50(1):49–58

Visentin M, Pagnoni A, Sarti E, Pietrogrande MC (2016) Urban $PM_{2.5}$ oxidative potential: Importance of chemical species and comparison of two spectrophotometric cell-free assays. Environ Pollut 219:72–79. https://doi.org/10.1016/j.envpol.2016.09.047

Wang H, Joseph JA (1999) Quantifying cellular oxidative stress by dichlorofluorescein assay using microplate reader. Free Radic Biol Med 27(5–6):612–616

Wang S, Guo Z, Xia P, Liu T, Wang J, Li S, Sun L, Lu J, Wen Q, Zhou M, Ma L, Ding X, Wang X, Yao X (2009) Internalization of NK cells into tumor cells requires ezrin and leads to programmed cell-in-cell death. Cell Res 19(12):1350–1362

Wang X, Xu Z, Su H, Ho HC, Song Y, Zheng H, Hossain MZ, Khan MA, Bogale D, Zhang H, Wei J, Cheng J (2021) Ambient particulate matter (PM_1, $PM_{2.5}$, PM_{10}) and childhood pneumonia: the smaller particle, the greater short-term impact? Sci Total Environ 772:145509

Weagle CL, Snider G, Li C, Van Donkelaar A, Philip S, Bissonnette P, Burke J, Jackson J, Latimer R, Stone E, Abboud I, Akoshile C, Anh NX, Brook JR, Cohen A, Dong J, Gibson MD, Griffith D, He KB, Holben BN, Kahn R, Keller CA, Kim JS, Lagrosas N, Lestari P, Khian YL, Liu Y, Marais EA, Vanderlei Martins J, Misra A, Muliane U, Pratiwi R, Quel EJ, Salam A, Segey L, Tripathi SN, Wang C, Zhang Q, Brauer M, Rudich Y, Martin R (2018) Global sources of fine particulate matter: Interpretation of $PM_{2.5}$ chemical composition observed by SPARTAN using a global chemical transport model. Environ Sci & Technol 52(20):11670–11681. https://doi.org/10.1021/acs.est.8b01658

Weichenthal S, Lavigne E, Evans G, Pollitt K, Burnett RT (2016) Ambient $PM_{2.5}$ and risk of emergency room visits for myocardial infarction: impact of regional $PM_{2.5}$ oxidative potential: a case-crossover study. Environ Health 15(1):1–9

World Health Organization. Air quality guideline (2006) Global update 2005. WHO regional Office for Europe, Copenhagen

Wu C, Yan Y, Chen X, Gong J, Guo Y, Zhao Y, Yang N, Dai J, Zhang F, Xiang H (2021) Short-term exposure to ambient air pollution and type 2 diabetes mortality: a population-based time series study. Environ Pollut 289:117886

Yang A, Jedynska A, Hellack B, Kooter I, Hoek G, Brunekreef B, Kuhlbusch TA, Cassee FR, Janssen NA (2014) Measurement of the oxidative potential of $PM_{2.5}$ and its constituents: the effect of extraction solvent and filter type. Atmos Environ 83:35–42

Yang Q, Chen Y, Krewski D, Shi Y, Burnett RT, Mcgrail KM (2004) Association between particulate air pollution and first hospital admission for childhood respiratory illness in Vancouver, Canada. Arch Environ Health: Int J 59(1):14–21

Yu L, Wang G, Zhang R, Zhang L, Song Y, Wu B, Li X, An K, Chu J (2013) Characterization and source apportionment of $PM_{2.5}$ in an urban environment in Beijing. Aerosol Air Qual Res 13(2):574–583

Zhang Z, Dong B, Li S, Chen G, Yang Z, Dong Y, Wang Z, Ma J, Guo Y (2019) Exposure to ambient particulate matter air pollution, blood pressure and hypertension in children and adolescents: a national cross-sectional study in China. Environ Int 128:103–108

Zhang Z, Guo C, Lau AK, Chan TC, Chuang YC, Lin C, Jiang WK, Yeoh EK, Tam T, Woo KS, Yan BP, Chang L-Y, Wong MCS, Lao XQ (2018) Long-term exposure to fine particulate matter, blood pressure, and incident hypertension in Taiwanese adults. Environ Health Perspect 126(1):017008

Zomer B, Collé L, Jedyńska A, Pasterkamp G, Kooter I, Bloemen H (2011) Chemiluminescent reductive acridinium triggering (CRAT)—Mechanism and applications. Anal Bioanal Chem 401(9):2945–2954

Seasonal Prediction and Evaluation of RSPM for an Open Cast Coal Mine

Navin Prasad and Tanushree Bhattacharya

1 Introduction

Most of the developing countries around the world rely on coal mining to meet their energy needs, which causes activities to produce a substantial amount of particulates. These activities are blasting, loading, drilling, hauling on unpaved roads, unloading, storing, etc. (Kumar et al., 2014; Lal & Tripathy, 2012; Pandey et al., 2014). Due to a better understanding of pollution sources and their levels, the significance of air pollution prevention has grown over the past several years. A major issue on global scale is chronic air pollution exposure. According to the WHO, air pollution is a factor in over 2.7 million deaths annually. Overall, particulate matter coming from coal mines frequently reaches hazardous levels, disrupting the environment and having a negative impact on human health (Sahu et al., 2018; Gautam et al., 2015). Chronic lung conditions like cancer, bronchial asthma, and chronic bronchitis are among the illnesses or health risks that are brought on by increased exposure to PM_{10} and $PM_{2.5}$ (USEPA Report, 2009; WHO, 2005).

Dispersion models now in use, like the AMS/EPA Regulatory Model (AERMOD), create model inputs by processing regularly accessible meteorological information. As a result, both the quantity and quality of meteorological data, and the choice of surface properties at the observing location, affect model predictions of concentrations (Isakov & Vlad et al., 2007; Seangkiatiyuth et al., 2011). Hourly surface and upper air meteorological readings that are stable and horizontally homogeneous are necessary for AERMOD (Kesarkar et al., 2007).

However, AERMOD's simulation of PM_{10} dispersion for complicated topography of open-pit coal mines or quarries as open-pit sources revealed a significant gap in the literature. A few research (Huertas et al., 2012; Tartakovsky et al., 2013) investigated AERMOD for mine or quarry complicated terrain. In AERMOD simulations of coal

N. Prasad · T. Bhattacharya (✉)
Civil and Environmental Engineering, Birla Institute of Technology, Mesra 835215, India
e-mail: tbhattacharya@bitmesra.ac.in

P. K. Padhy et al. (eds.), *Airborne Particulate Matter*,
https://doi.org/10.1007/978-3-031-84408-9_10

mines for dispersion or deposition, the geomorphological component also plays a crucial role in overall calculation and modelling. In contrast, open-cast coal mining operations significantly alter the topography of the ground in order to do excavation work. As a result, the pit shape and aspect ratio—the ratio of mine length (L) to depth (D)—has a major impact on wind flow. The depth and width of the terrain variations range between 200 and 300 meters. As open-cast coal mines expand, they tend to become both more productive and deeper (Peng et al., 1995). The AERMOD model must be parameterized in real time to account for the increasing PM_{10} emission in order to manage particle dispersion in the pit and off-site locations. Complex terrain produces local-scale circulations (Srivastava et al., 2021).

Understanding the behaviour of $PM_{2.5}$ and PM_{10} emissions from open pit mines and the impact of meteorological dynamics on $PM_{2.5}$ and PM_{10} concentration is the main aim of this study. To comprehend the effects of the particulate matter dispersion from the mine and moving vehicles on the surrounding environment, AERMOD was utilized to simulate this dispersion. To comprehend the effects of numerous operations in a neighbouring residential area and the effect of wind speed and wind direction on the overall dispersion, spatial monitoring was also carried out. The results of the study will be useful in creating appropriate defences against air pollution from the coal mining area.

2 Methodology

2.1 Study Area

The 8.28 km^2 Khadia Open Cast coal mine selected for this study is situated in Singrauli Coalfield and comes under Northern Coalfield Limited producing 16% of the Country's coal and meeting 10% of India's total energy needs. There are nine open coal mines running simultaneously with four power plants working in the periphery of these mines. The ecosystems and community are highly affected because of the high density of emission sources, and the community also complains about the area's poor air quality. Tropical deciduous forests were heavily populated in the area prior to coal mining and other industrial activity. The region's terrain has undergone tremendous modification as a result of large-scale coal mining operations. In addition to coal mining, the nearby industrialization had a negative impact on the area's air and water quality, land use and cover. In Singrauli open-cast mines, 12.2 tons of dust are produced per day, with 0.18 tons of dust emitted due to blasting and drilling operations. Land cleaning, which involves using dozers and scrapers to remove topsoil and maintain soil quality and plant communities, is the first step in a mining operation. Figure 1 shows the study area with the CAAQMS station installed in the mine area.

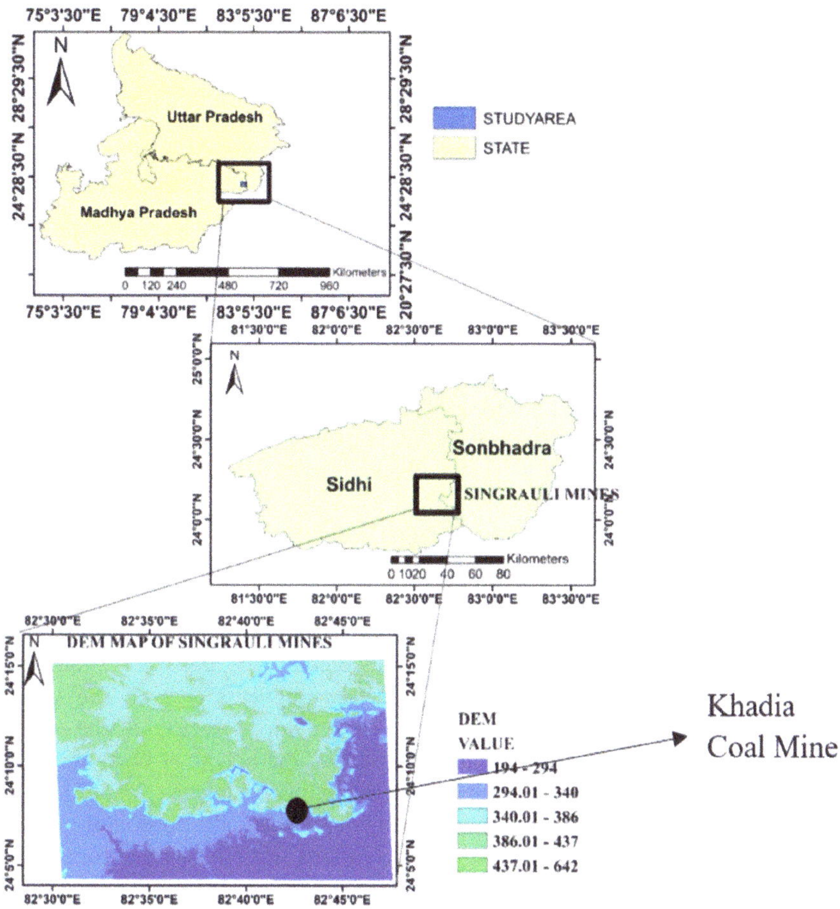

Figure 1: Study area with CAAQMS station.

The main pollution source in the open-cast mine are the vehicles moving on the unpaved haul road. The loading, unloading, drilling, shoveling, and stockpiling process also contribute a considerable amount to the overall pollution load.

2.2 *Meteorological Data and Dispersion Model*

The 2020-2021 meteorological data used in this study were collected from meteorological stations. The hourly surface data were organized into the .xls format using Microsoft Excel 2019. Every 15 min interval data of the entire year were converted into 24 hours data. The wind speed, wind direction, ambient temperature, cloud cover, and ceiling height were all included in the hourly surface data. It was measured at the meteorological station in Khadia open-cast coal mines at coordinates 24°07′20.00″ N and 82°41′04.20″ E. Studying four major ambient air contaminants, including particulate matter smaller than 10 microns (PM_{10} and $PM_{2.5}$), sulfur dioxide (SO_2) and oxide of nitrogen (NO_x).

AERMOD has been used to investigate simulating air quality. Essentially, the AERMOD is a Gaussian dispersion model with the boundary layer theory integrated along with terrain characteristics, turbulence and dispersion. The air quality model requires two types of data inputs: first, meteorological information, which is pre-processed using AERMET software, and second, emission rate data from various activities within the mining area. The raw meteorological data was formatted according to the specifications for the AERMOD model: upper air data and hourly surface data were organized into the SCRAM format that was set up in the FSL format. The data arrangement produced the SFC File (*. SFC) forms, which is basically a surface file and PFL File (*.PFL), etc. Figure 2 shows the input required to create the Samson file, which is used as input to AIRMET to create the SFC file format.

Following are inputs of the Samson file which include wind speed, relative humidity, dry bulb temperature, opaque cloud cover, wind direction, ceiling height, hourly precipitation amount, global horizontal radiation, year, month, day and hour data. Once the Samson file is created it is entered as an input file in AERMET View with other parameters like base elevation, latitude, longitude, upper air parameters, land use type like Albedo, Bowen Ratio, and surface roughness. (Fig. 3). The Albedo for the given site has been taken as 0.29, the Bowen ratio as 0.425, and the surface roughness as 0.04025. Surface roughness describes the height of obstructions to wind flow, and its length plays a crucial role in regulating the intensity of mechanical turbulence and the stability of the boundary layer. Albedo is the portion of incident solar light that is completely reflected by the surface without being absorbed. An

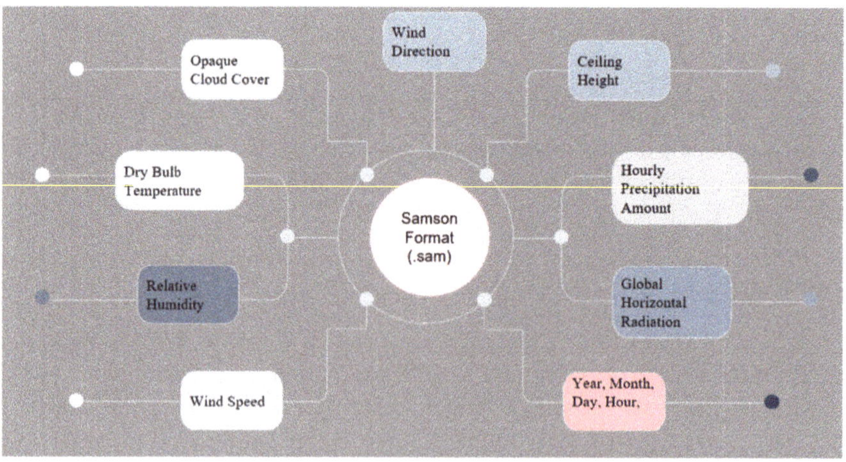

Figure 2: Input required to make Samson file.

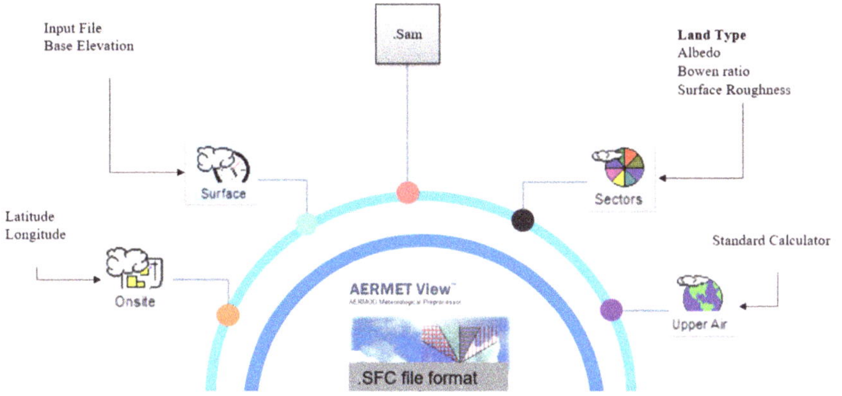

Figure 3: Input required in AERMET.

indication of surface moisture is the daylight Bowen Ratio (ratio of sensible heat flux to latent heat flux). To calculate how much incoming radiation is converted to sensible heat flow, the Albedo and Bowen Ratios are utilized.

Meteorological parameters from the created .SFC file is used as input. The area source has been chosen as the Khadia mines' primary source of pollution. The latitude and longitude of the CAAQMS station have been used as the central point of reference for the 10×10 sq. km by 500 m cartesian area grid. Using AERMAP as a terrain data preprocessor, complicated terrain is incorporated using Digital Elevation Data. To obtain a roughly good result for the modelling, SRTM1 (Shuttle Radar Topography Mission) has been utilized with 30 m resolution as terrain elevation data. The control option in AERMOD is used to choose the kind of pollution deposition and the kind of pollution with the dispersion coefficient (either rural or urban). The model's projection of the data into Google Earth allows users to examine the model's exact position and pollution distribution for every zone, including the core and buffer zone of open cast coal mines. The various input required is mentioned in Fig. 4.

2.3 *Emission Rate Calculation*

There were numerous open-pit sources close to the Khadia mines. The emission rates from coal mines' open surfaces were evaluated as an open pit source, whereas coal dumps and OB dumps were regarded as an area source. Table 1 provides the emission rate calculation equation for Indian mining circumstances. It takes into account four key inputs: the leasehold area, production in million metric tons, wind speed in meters per second, and overburden in cubic meters. The equation used to calculate the emission rate has been taken from the research paper "Development of empirical formulae to determine emission rate from various open-cast coal mining

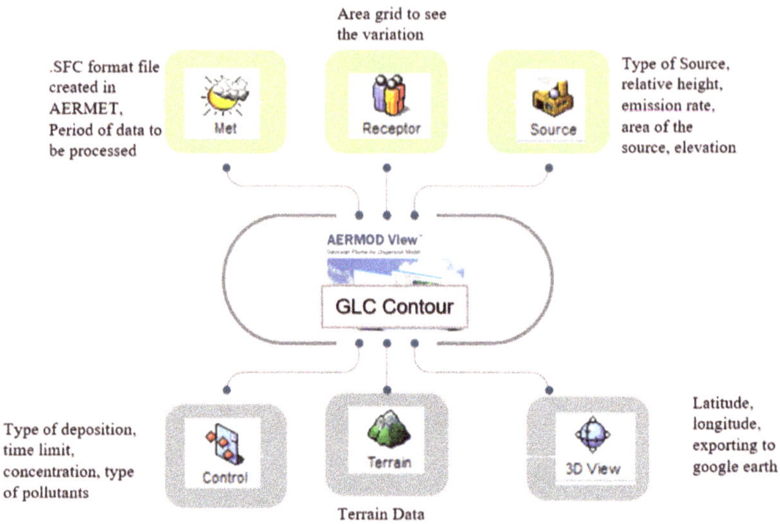

Figure 4: Input required in AERMOD.

operations" (Chaulya et al., 2002). Table 1 lists the various parameters needed to calculate the emission rates, while Table 2 presents the emission rates in g/s for all four pollutants.

From January to December 2020, field measurements were made to track the concentration of the pollutant. Each location's average for the entire season was taken. The stations were also placed in nearby villages located in buffer zones (10 km away from the mines). fine dust sampler (FDS) and respiratory dust samplers (RDS) were used on-site to conduct the monitoring. Air flowed through a Whatman glass fiber filter at a rate of 1.1 m^3 per min accumulating PM_{10} on it. Using the gravimetric method, the concentration of particulate matter is determined.

Table 1: Parameters required to calculate emission rate

Unit	*m/s*	*Mt/year*	*Mm^3/year*	*km^2*
Parameters	(u) WS	(c) CP	(b) OB	(a) LA
Khadia	0.836	14	60.58	16.4

Table 2: Emission rate with unit and pollutants

Pollutants	*PM_{10}*	*$PM_{2.5}$*	*NO_2*	*SO_2*
Unit	g/s	g/s	g/s	g/s
Khadia	15.36	9.98	1.78	1.13

3 Result

In summer, meteorological parameters such as temperature increase, whereas wind speed and humidity decrease causing a reduction in all pollutants in the year 2020 (Tartaglione, 2020). Monsoon rainfall causes temperature reduction and a rise in humidity resulting in the least ambient air pollution concentration compared to other seasons. Once monsoon ends, the weather starts to get drier. During this period, the concentration of the pollutant starts to rise again with the highest concentration in winter, followed by post-monsoon.

Figure 5 depicts the concentration of PM_{10} and Fig. 6 for $PM_{2.5}$ for all the seasons throughout the year. The variation in concentration is moderate in magnitude till the onset of the pre-monsoon season. Warm air is faster and less dense than cold air. Cold air traps the pollution. In the absence of sunlight, it has a longer lifetime in the atmosphere, which is the reason for high SPM in winter, but some days in the summer have high concentration as compared to winter because high production levels and low mitigation measures make the dust spread more into the atmosphere and pollute the environment severely.

Data from the meteorological station at the Khadia mines have been compared to the wind rose in the plot. The wind vectors were plotted using the AERMOD preprocessor AERMET. The four seasons' variations in wind direction are shown

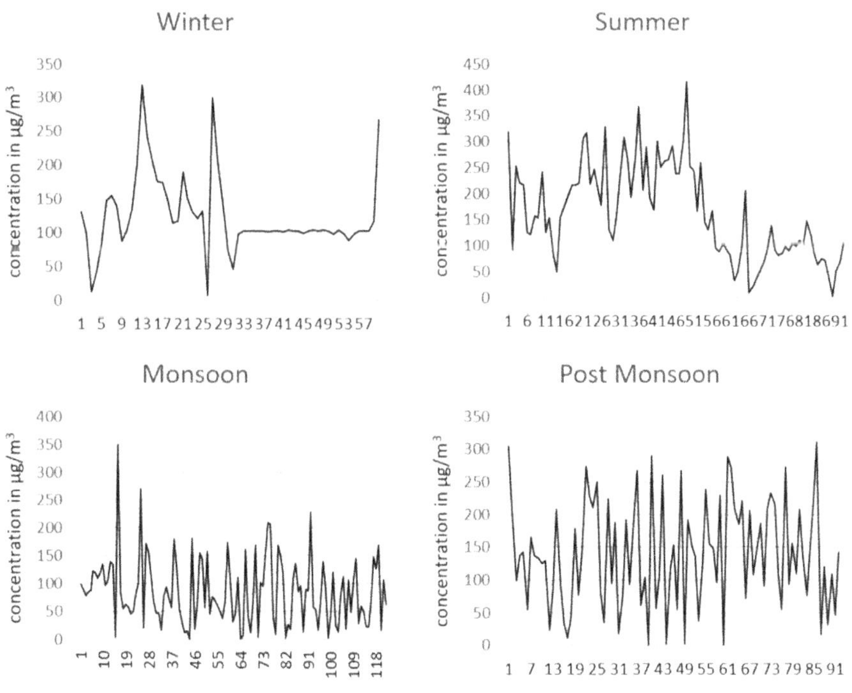

Figure 5: The PM_{10} seasonal variation of winter, summer, post-monsoon, monsoon (clockwise).

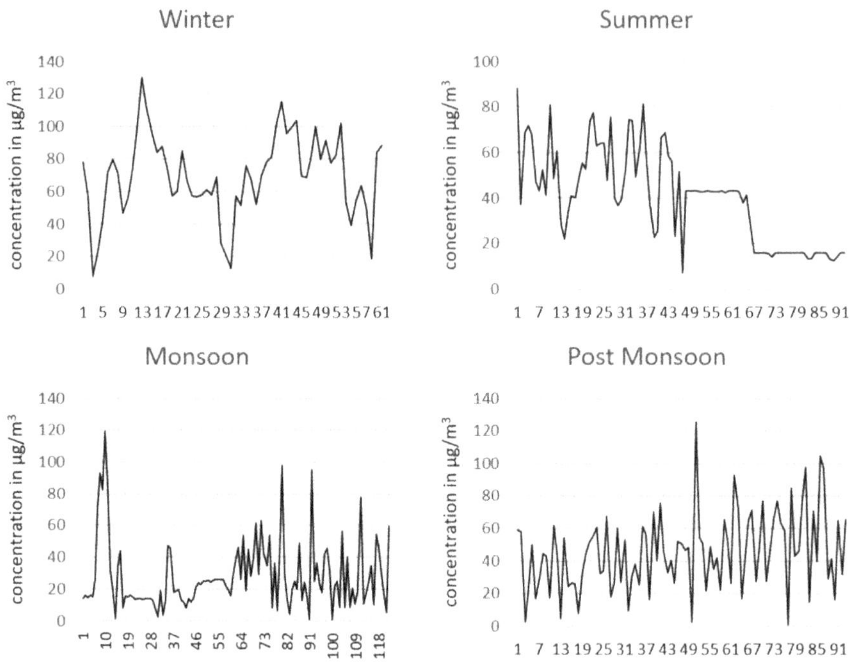

Figure 6: The $PM_{2.5}$ seasonal variation of winter, summer, post-monsoon and monsoon (clockwise).

in Fig. 7. The predominant wind direction is the west-south-west (WSW), although there may be periodic variations of wind from the southwest.

Figure 8 shows the annual output of AERMOD; the bright contours define the concentration of PM_{10} and $PM_{2.5}$ near the core mining site and 10 km away from the Khadia mines as a buffer zone. The area far from the mining site is less affected, but due to wind direction, the area in the west direction is severely affected since the prevailing wind is also found in the same direction throughout the year, there is less of an impact in the area far from the mining site (Fig. 7). Some of the contaminants are dispersed above the Dudhichua mines, which are situated to the area's north. The concentration is found to be 310 micrograms per cubic meter, which is higher than the center area of the Khadia mining site. The mining region of Khadia is indicated by the contour in red. The primary site area's northern and eastern directions have the greatest dispersion. The AERMAP aids in predicting the impact of the terrain.

Table 3 describes the seasonal result of particulate matters in all the four seasons and also the monitoring values from the CAAQMS (Central Ambient Air Quality Monitoring Station) installed in Khadia mine. All three columns' values are highest in the AERMOD column. Winter and summer require the most concentration. When compared to the AERMOD, the monitored levels are significantly lower, but they are greater than the CAAQMS value. The prevailing wind direction is what causes the difference. Due to the CAAQMS station's installation on the upwind side, it captures

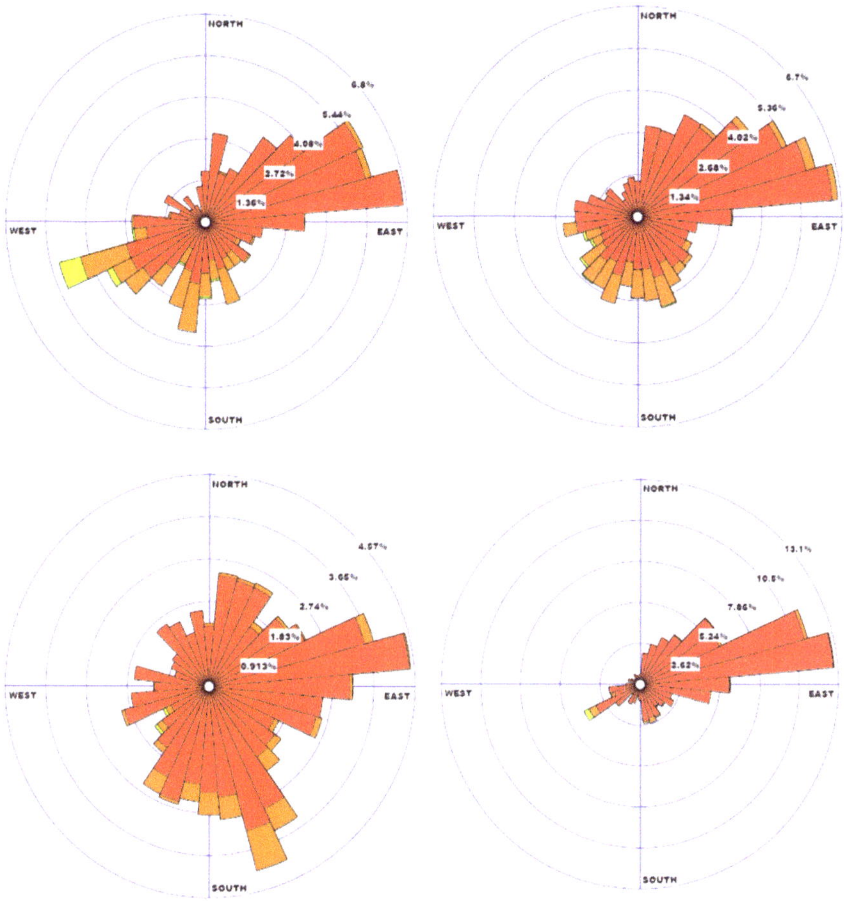

Figure 7: Wind rose for all the seasons: winter, summer, post-monsoon and monsoon (clockwise).

very little pollution, whereas monitoring and the mine are downwind side, which faces northwest, has the highest concentrations. Because of the southeast wind in the winter and after the monsoon, some pollutants are captured by the CAAQMS, while during rest of the year, the direction in which pollutants spread is different. The location chosen for the monitoring must be both upwind and downwind so that the comparison of contaminants may be done quickly and accurately. Khadia open cast project's community center and market were chosen as the monitoring site nearby the CAAQMS station, which is in a downwind direction; this is why the concentration of pollutants is higher than it is in the case of CAAQMS.

It is important to remember that the discussion was based on maximum concentrations that occur in specific areas. Figure 8 illustrates how simulated concentrations are typically lower than conventional values in most places. Typically, the major

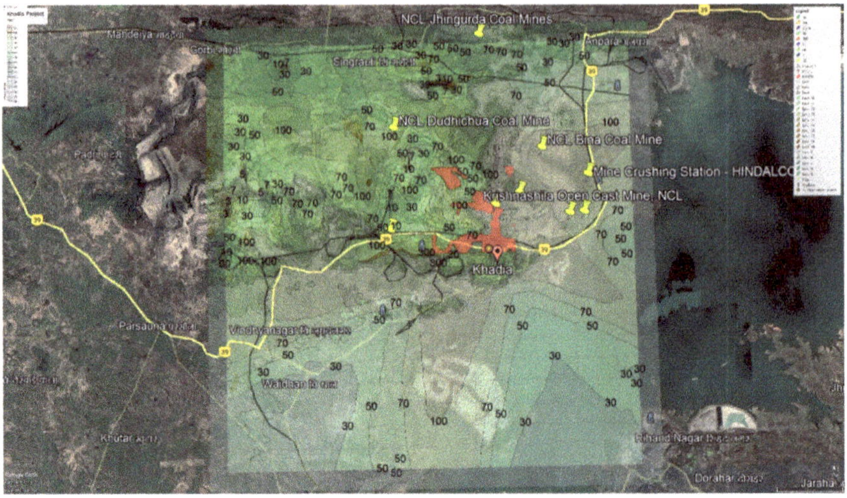

Figure 8: PM_{10} maximum concentration contours for the given AERMOD averaging times (annual) placed on Google Earth.

Table 3: Emission rate with unit and pollutants

Seasons	*Khadia*			
	Pollutants	*AERMOD*	*Monitored*	*CAAQMS*
Winter	PM_{10}	310	195	112.23
	$PM_{2.5}$	147	37	63.49
Summer	PM_{10}	295	159	157
	$PM_{2.5}$	164	46	35.8
Monsoon	PM_{10}	198	47	68.32
	$PM_{2.5}$	100	24	23.22
Post monsoon	PM_{10}	200	193	125.97
	$PM_{2.5}$	100	72	36.54

The concentration of Pollutants is in $\mu g/m^3$

goal of this kind of research is to identify high-risk areas (hot spots) for additional investigation.

4 Conclusion

In this article, according to mining standards, pollutants have high concentrations in the Khadia coal mines in the summer of 2020. Monsoon rainfall causes temperature reduction and a rise in humidity resulting in the least ambient air pollution concentration compared to other seasons. Once the monsoon ends the weather starts to get drier. During post-monsoon the concentration of the pollutant starts to rise again. Cold air traps the pollution in the absence of sunlight. PM_{10} and $PM_{2.5}$ high concentrations were found in winter due to cool and dry weather. The dispersion pattern is considerably impacted by elevation, according to the isopleths of all simulations. The wind pattern is extremely important for successful and accurate modeling estimations, and model predictions were both extremely cautious and overstated. The findings also helped mine workers understand how wind direction affects concentration fluctuation. The model's limitations (at benches of coal quarries, wind flow causes pit retention and recirculation) need to be sufficiently overcome in order to determine the health risks to miners. The inventory of emission components and model findings for PM_{10} dispersion modeling also made it feasible to understand the movement of PM_{10} in the difficult topography of coal mines.

References

Chaulya SK et al (2002) Development of empirical formulae to determine emission rate from various opencast coal mining operations. Water Air Soil Pollut 140:21–55

Gautam S, Prusty BK, Patra AK (2015) Dispersion of respirable particles from the workplace in opencast iron ore mines. Environ Technol Innov 4:137–149

Huertas JI, Huertas ME, Izquierdo S, Gonzalez ED (2012) Air quality impact assessment of multiple open pit coal mines in northern Colombia. J Environ Manag 93:121–129

Isakov V et al (2007) Evaluating the use of outputs from comprehensive meteorological models in air quality modeling applications. Atmos Environ 41:1689–1705

Kesarkar AP, Dalvi M, Kaginalkar A, Ojha A (2007) Coupling of the weather research and forecasting model with AERMOD for pollutant dispersion modeling. A case study for PM_{10} dispersion over Pune, India. Atmos Environ 41:1976–1988

Kumar P, Morawska L, Birmili W, Paasonen P, Hu M, Kulmala M et al (2014) Ultrafine particles in cities. Environ Int 66:1–10

Lal B, Tripathy SS (2012) Prediction of dust concentration in open cast coal mine using artificial neural network. Atmos Pollut Res 3:211–218

Pandey B, Agrawal M, Singh S (2014) Assessment of air pollution around coal mining area: Emphasizing on spatial distributions, seasonal variations and heavy metals, using cluster and principal component analysis. Atmos Pollut Res 5:79–86. https://doi.org/10.5094/APR.2014.010

Peng X, Lu GR (1995) Physical modeling of natural wind and its guide in a large open pit mine. J Wind Eng Ind Aerodyn 54–55:473–481

Sahu SP, Patra AK, Kolluru SSR (2018) Spatial and temporal variation of respirable particles around a surface coal mine in India. Atmos Pollut Res 9(4):662–679. https://doi.org/10.1016/j.apr.2018.01.010

Seangkiatiyuth K, Surapipith V, Tantrakarnapa K, Lothongkum AW (2011) Application of the AERMOD modeling system for environmental impact assessment of NO_2 emissions from a cement complex. J Environ Sci 23(6):931–940

Srivastava A, Kumar A, Elumalai SP (2021) Evaluating dispersion modeling of inhalable particulates (PM_{10}) emissions in complex terrain of coal mines. Environ Model Assess 26(3):385–403
Tartaglione N et al (2020) A note on the statistical evidence for an influence of geomagnetic activity on Northern Hemisphere seasonal-mean stratospheric temperatures using the Japanese 55-year Reanalysis. Annales Geophysicae 38(2). Copernicus GmbH
Tartakovsky D, Broday DM, Stern E (2013) Evaluation of AERMOD and CALPUFF for predicting ambient concentrations of total suspended particulate matter (TSP) emissions from a quarry in complex terrain. Environ Pollut 179:138–145
USEPA (2009) Integrated science assessment for particulate matter. National Center for Environmental Assessment. RTP Division
WHO (2005). http://apps.who.int/iris/bitstream/10665/69477/1/WHO_SDE_PHE_OEH06.02_eng.pdf

Seasonal Health Risk Assessment and Prediction of $PM_{2.5}$ Concentrations with a Gender Bias in Exposure after Wearing Masks

Monalin Mishra, Trupti Das, and Boopathy Ramasamy

1 Introduction

Population health research has been undertaken over the last decade to investigate the relationship between regional ambient particulate levels, and health tends to reveal that prevailing pollution levels in the air constitute a major local, regional, national, international, and perhaps even worldwide concern for public health. According to the WHO, air pollution is the leading environmental health risk worldwide, causing approx. 7.1 million premature deaths each year, 4.2 million of which are linked to outdoor air pollutants (https://www.who.int/health-topics/air-pollution#tab=tab_1). There is significant evidence that both short- and long-exposures are linked to premature death and a variety of respiratory and cardiovascular disorders (Chatignoux et al., 2014). Polluted air is a complicated mixture of particulates and gaseous co-pollutants released by both local and remote sources. Majority of recent research findings and policy development efforts have focused on the health risks of ambient particulate matter (PM), and while delegates acknowledged the importance of gaseous co-pollutants and toxic gases, particulates were the prominent topic of discussion (Craig et al., 2007). Particulates are a complex mixture whose both chemical and physical constitutions change over time and appear to be influenced by climatic and long-term flight circumstances as well as unique source factors. However, there is now a growing body of research linking $PM_{2.5}$ to negative health outcomes (Shrivastava et al., 2017). The public health impacts of minute $PM_{2.5}$ have prompted numerous nations to implement specific air quality regulations or norms. This has resulted in a demand for an understanding of the primary sources of $PM_{2.5}$. Over the past decade, epidemiological data have revealed that exposure to air pollutants at the

M. Mishra · T. Das (✉) · B. Ramasamy
Aerosol & Trace gases Laboratory, Environment & Sustainability Department, CSIR-Institute of Minerals & Materials Technology (CSIR-IMMT), Bhubaneswar, Odisha 751013, India
e-mail: truptidas@immt.res.in

Academy of Scientific and Innovative Research (AcSIR), Ghaziabad 201002, India

P. K. Padhy et al. (eds.), *Airborne Particulate Matter*,
https://doi.org/10.1007/978-3-031-84408-9_11

existing pace in metropolitan is correlated with a rise in mortality and a number of health issues. Because particulates from local emissions are placed on the regional backdrop, average $PM_{2.5}$ levels are greater in urban areas. While urban activities undoubtedly contribute to $PM_{2.5}$ levels, a significant portion of the $PM_{2.5}$ recorded in cities comes from sources upwind. To limit the danger to health and the environment, air quality monitoring plans must include details on the relative influence of local versus upwind sources, as well as the main sources of $PM_{2.5}$ (Chatignoux et al., 2014). Thus, estimating the overall health risks of air particles quantitatively can lead to early preventive interventions and enhanced health standards. There have been numerous publications (Kim et al., 2020; Tao et al., 2014; Hansen et al., 2012 and a lot more on the increase in death rate owing to PM exposure as a pollutant that may infiltrate the respiratory system of the human.

Since late 1990's, ongoing research initiatives on PM pollution analysis reports in Bhubaneswar (India) have yielded significant scientific results (Mahapatra et al., 2018), for example, published three years of PM_{10} and $PM_{2.5}$ observations, as well as seasonal variations in meteorological conditions. However, in-depth research on the origin, dispersion, and effects of ambient aerosols on human health is essential to disprove key data that would be helpful to the scientific community in understanding the overall effects of PM health risk (Panda et al., 2021b). Furthermore, risk evaluation of particular constituents in ambient aerosols is critical in terms of air pollution-related health problems in this location.

2 Study Area

Long-term PM2.5 measurements for 6 years (2016 to 2021) have been taken at the CSIR-IMMT in Bhubaneswar (India) in a traffic-oriented location and an urban backdrop site in the city centre. Based on the distinctive weather conditions, shifting meteorological characteristics and primarily on monsoon occurrences, the weather of Bhubaneswar (India) is split into four distinct seasons: pre-monsoon (March, April and May), monsoon (June, July and August), post-monsoon (September, October and November), and winter (December, January and February) (Zeb et al., 2019; Panda et al., 2021b). During winter, the pollutant had higher daily mean values. The cleanest month on average is July, when the amount of precipitation is usually considerable and there is a washout phenomenon. December is the most polluted month, with mean readings 3–4 times higher than the average values in July. The greater number in winter may reflect an increase in particle occurrence due to both nucleation and accumulation mechanisms, as well as poor atmospheric blending characteristics. The city receives most solar radiation during pre-monsoon months, when temperatures range from 40 to 45°C. At the same time, in winter, the minimal temperature drops to as low as 10°C. During monsoon, the location receives a lot of precipitation, with a monthly average of 220 mm.

3 Methodology

The $PM_{2.5}$ measuring procedure is an essential issue in terms of a nationally standardised protocol for accomplishment assessment. It is commonly understood that new technologies can result in varying concentration levels, and that $PM_{2.5}$ is described by how it is analyzed (Panda et al., 2021b). A low volume sampler (Envirotech, 550 MFC) was used to collect the 24-hour $PM_{2.5}$ samples on the roof of CSIR-IMMT at flow rates of 1 ms^{-1} and 16.7 LPM, respectively. The pre-baked Whatman quartz microfiber filter papers for $PM_{2.5}$ were used to collect the samples. Sampling was carried out twice a week every month from 2016 to 2021. A Sartorius weighing scale with 0.00001 mg sensitivity was used to determine the filter paper's initial and final weights (Nath et al., 2021). The gravimetric approach was used to calculate $PM_{2.5}$ concentrations by subtracting the initial weight of the filter sample from the final weight and then dividing it by the volume of air travelled through it. Before sampling, filters were pre-conditioned in desiccators at room temperature for 24-hours to achieve consistent humidity (Flores et al., 2020). The instrumentation and operational mechanisms of the sampler were followed as per USEPA guidelines, and the same methods have been discussed in detail earlier (Panda et al., 2016).

3.1 Risk Assessment

The excess risk associated with pollutant loads was used to assess the possible health benefits in different seasons due to changes in concentrations (Sharma et al., 2020). Following equation is used to calculate the relative risks of a pollutant

$$RR_i = \exp\left[\beta_i (C_i - C_i,\ 0)\right] \tag{1}$$

where $C_i > C_{i,0}$, RR_i = relative risk of a pollutant (i), β_i = coefficient of exposure-response indicating the added health risk i.e. death caused by per unit of air pollutant, when it crosses a threshold level whose values is given for $PM_{2.5}$ = 0.038% per μgm^{-3} (Hu et al., 2015; Sharma et al., 2020; Shen et al., 2020) and $C_{i,0}$ = threshold concentration meaning that when the concentration of the pollutant (i) is below or equal then it has no ER.

The excess health risk (ER) for pollutant (i) can be estimated using the formulae

$$ER_i = RR_{i-1} \tag{2}$$

3.2 Prediction Model

This study focuses on air pollution forecasting in the Bhubaneswar region (India) utilising $PM_{2.5}$ mass concentration data for all four seasons. The primary goal of this study is to predict $PM_{2.5}$ concentrations based on past concentration trends. To forecast $PM_{2.5}$ concentration in the Bhubaneswar study area, the study employs time series analysis and linear regression. According to findings, if suitable mitigation measures are not implemented, air quality in Bhubaneswar region will deteriorate.

The $PM_{2.5}$ data obtained over the years was pre-processed and translated to monthly data. Using time series analysis and linear regression, the processed data was utilised to forecast concentration. The standard multiplicative model for time series analysis is as follows:

$$Y_t = S_t \times I_t \times T_t \tag{3}$$

where Y_t = predicted data, S_t = Seasonal factor, I_t = Irregularity factor and T_t = trend factor.

The training data was de-seasonalized (S_t), irregularity (I_t) was eliminated, and the trend factor (T_t) was computed using linear regression utilising the centre moving average (CMA) (Iii and Iii, 2022). Future concentrations were computed using equation (3) for the years 2022-2025 (Fig. 3).

3.3 Exposure Dosage (d)

Following equation was used to obtain the average lifespan of daily exposure dosage (ALDED):

$$\text{ALDED} = \frac{C \times IR \times EF \times ED \times ET}{H \times BW \times AT} \tag{4}$$

where ALDED is the average lifespan of daily exposure dosage [$\mu g\ m^{-3}$ (kg day)$^{-1}$] (Yan et al. 2021), IR the inhaled rate of breathing [m^3 day^{-1}], for male, MIR = 18.65 and for female, FIR = 14.80 (Yan et al., 2021), EF the exposure frequency, i.e., the number of work days for each scenario [day a^{-1} or D a^{-1}] = 60 days per 24 years for both males and females (Panda et al., 2021b), ED the exposure duration, i.e., the number of years in career for each scenario [a] = 24 years for both males and females (Panda et al., 2021b), ET the daily exposure time [hours] = 24 hours, H the number of hours in a day [hours] = 24 hours, BW the body weight [kg], for male (MBW = 68.80 kg and for female, FMW = 58.95 kg) (Yan et al., 2021) and AT the average exposure time [lifetime] [day] = ED × 365 days year^{-1} = 24 × 365 days year^{-1} (Panda et al. 2021b).

Equation (5) was used to calculate the exposure dose (*d*) [$\mu g\ m^{-3}$ (kg day)$^{-1}$]

$$d = (1 - \alpha) \times ALDED \tag{5}$$

where α represents the parameter of the model equation, for people without wearing a mask, $\alpha = 0$ and for people wearing a mask, $\alpha = 0.95$ (Yan et al., 2021).

4 Result and Discussion

Table 1 gives descriptive statistical data obtained from the entire set of 24-h average $PM_{2.5}$ values recorded at the institution site from January 2016 to December 2021. During all six years, the ambient air mass concentration of $PM_{2.5}$ ranged from 10 to 153 μgm^{-3}, with an annual average of 57.52 μgm^{-3} in 2016, 62.01 μgm^{-3} in 2017, 54.31 μgm^{-3} in 2018, 60.85 μgm^{-3} in 2019, 47 μgm^{-3}

in 2020, and 45.39 μgm^{-3} in 2021. During all years, the seasonal fluctuation of ambient $PM_{2.5}$ mass concentration follows a trend, with the greatest mass concentrations reported during the winter months (76.69 to 107.8 μgm^{-3}), followed by the post-monsoon (43.79 to 68.4 $\mu g/m^3$) and pre-monsoon (21.33 to 45.11 μgm^{-3}). For all the years, lowest ambient $PM_{2.5}$ concentration was reported during the monsoon season (18 to 45.11 μgm^{-3}).

Figure 1 shows seasonal fluctuations in ambient $PM_{2.5}$ mass concentrations for all years. The difference in wind patterns is one of the major factors for the shift in seasonal trends. Higher wind speeds during the pre-monsoon and monsoon seasons aid in the diluting of ambient air pollutants, but stagnant atmospheric conditions and lower wind speeds during winter and post-monsoon seasons allow ambient $PM_{2.5}$ concentrations to build up (Mahapatra et al., 2014, 2018). Rainfall, relative humidity (RH), temperature and boundary-layer conditions, in addition to wind speed and direction, have a substantial effect on the fluctuating ambient $PM_{2.5}$ mass concentrations. The results of the relative health risk assessment during different seasons are given in Fig. 2.

It was discovered that the risk was greatest for all years when exposed during the winter months, particularly during winter of 2021. During monsoon, the risks were minimal, but in surprise, there was no mean excessive danger from $PM_{2.5}$ during the summers of 2020 and 2021, and monsoon of 2020 since full or partial limited activities were applied during lockdowns. While significant health hazards were

Table 1 Descriptive statistical data obtained of 24-hours average $PM_{2.5}$ concentration (in μgm^{-3}) recorded at the site from 2016 to 2021

	2016	*2017*	*2018*	*2019*	*2020*	*2021*
Mean	57.52 ± 7.58	62.01 ± 7.44	54.31 ± 8.96	60.85 ± 10.90	47.00 ± 11.56	45.39 ± 9.28
Minimum	30.85	40.34	21.86	25.42	13.94	10.22
Maximum	102.36	116.04	125.27	153.01	139.55	109.55

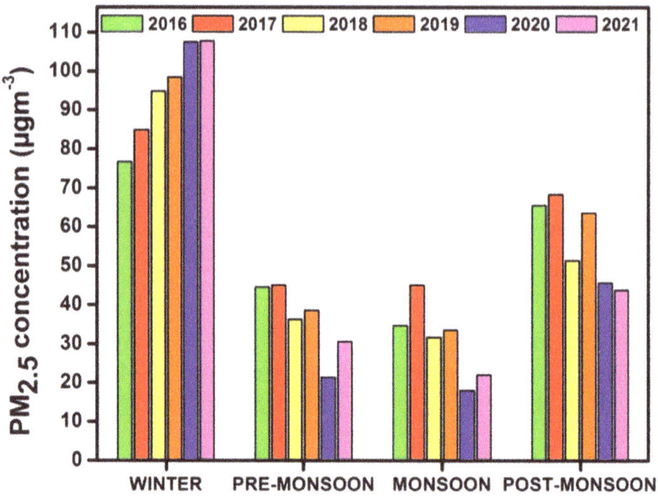

Figure 1: Seasonal variation of $PM_{2.5}$ from 2016-2021.

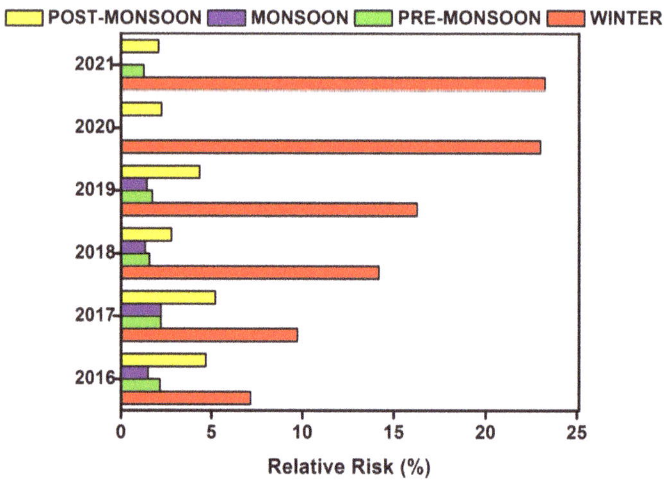

Figure 2: Seasonal relative health risk assessment of $PM_{2.5}$ during 2016–2021.

attained during the lockdown, the excessive risks (ER) calculated from the criteria pollutant also lowered due to blockage of principal sources of emission, indicating the need of focusing on source profile minimization. Particulate matter (PM) revealed different behaviour throughout the Covid-19 control period compared to the reference period, with mean concentrations decreasing (from 41.13 to 37.18 μgm^{-3}) during the Covid period of 2020. The greater reduction in $PM_{2.5}$ could be attributed to its greater contribution from anthropogenic causes (Klimont et al., 2017). The primary

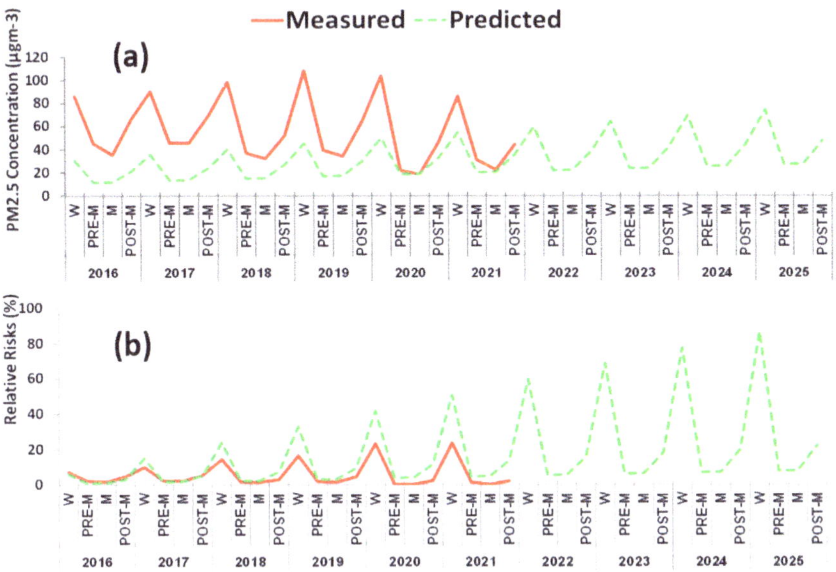

Figure 3: Seasonal variation of the original and the predicted: (a) $PM_{2.5}$ concentration and (b) relative risk (RR) from 2016-2025 (W = Winter, PRE-M = Pre-monsoon, M = Monsoon, POST-M = Post-monsoon)

sources of $PM_{2.5}$ in urban environment are road traffic, fuel combustion, industrial sources, construction work, dust resuspension and regional air mass transport, which may have an impact on PM to the point of decreasing impacts of local emission reduction. Aerosol particles $PM_{2.5}$ are also garnering worldwide attention, as they have the potential to have a bigger negative impact on human health, particularly in metropolitan areas influenced by heavy traffic density or industrialization (Dongarrà et al., 2010)

Figure 3 shows the original and predicted $PM_{2.5}$ concentration and the corresponding relative risk (RR) associated with the pollution during various seasons for the preceding six years and the prediction for next four years. According to WHO air quality guidelines, the $PM_{2.5}$ threshold value of 25 μgm^{-3} (24-hour mean) was used for computation. Overall, significant health concerns from $PM_{2.5}$ were found in all seasons of every year, except during the lockdown time in 2020 and 2021. The statistical prediction for the $PM_{2.5}$ health risk plot from 2016 to 2025 revealed a disrupted trend caused by sudden suppressed sources of emissions as a result of shutdown enforcement on everyday and economic operations in the midst of COVID-19 epidemic. WHO amended the criteria for air quality requirements in 2021. Furthermore, during the nationwide lockdown, the air quality in the research area significantly improved (Nath et al., 2021; Panda et al., 2021a).

Eventually, as the unlock phases began, air quality began to deteriorate again. The reduced pollution concentrations during lockdown demonstrate that our planet's ecology can regenerate itself swiftly. Pollution can be decreased if proper and consistent actions are taken. This overall decline in RR in correspondence to the predicted value highlights the fact that the lockdown period could save a million lives each year from PM pollution impact. Monitoring and mitigating pollution in the research region are critical. The goal of this study is to forecast the concentration of $PM_{2.5}$ in business-as-usual situations. This research will assist policymakers in implementing pollutant concentration reduction measures in the study area and related demographic and geographic locations. According to the findings, the concentration of $PM_{2.5}$ under monitoring will continue to climb under the current business-as-usual situation. Furthermore, the study demonstrates that the bulk of PM is caused by anthropogenic activities such as biomass burning.

Figure 4 depicts the exposure dose of males and females separately from 2016 to 2021 during different seasons. Though the exposure pattern for both males and females followed the same seasonal trend being proportional to the PM concentration level with high exposure dose during the winter and lowest during monsoon, however, male exposure doses (d_m) were consistently higher than female exposure doses (d_f) for each exposure condition, despite being of the identical magnitude level. This resulted from males having a significantly higher inhalation breathing rate than females. The dosage response was mostly driven by the interaction of $PM_{2.5}$ exposure concentration with inhaled breathing rate (IR) and exposure time (ET) (Jahne et al., 2015; Hasan et al., 2019). As a result, the male exposure percentage was higher than the female exposure concentration during the same exposure time.

Furthermore, Fig. 5 shows around 95% reduction in the exposure dose for both males and females for all the years after wearing a mask. Even though both were of the same order of magnitude, the frequency of exposure risk of persons who did not wear masks surpassed the threshold. However, based on a conservative approximation, their risk of infection was one order of magnitude greater than the standard. When using a mask, the risk of infection was one order of magnitude lesser than when not wearing. The likelihood of infection risk can generally be reduced to a level that is tolerable for the general public by wearing a mask. As a result, wearing a mask is a useful protective precaution for those who have been exposed (Yan et al., 2021). This finding demonstrates that the illness health burden for exposed populations without masks is much higher, yet using masks can minimise the risk of disease health burden.

5 Conclusion

According to the findings, $PM_{2.5}$ concentration information cannot be extrapolated considering only these markers; hence, independent measuring tools are necessary to investigate the links between healthcare outcomes and PM concentration. The statistical prediction for the $PM_{2.5}$ health risk plot from 2016 to 2025 revealed a disrupted trend caused by sudden suppressed sources of emissions as a result of shutdown enforcement on everyday and economic operations in the midst of the COVID-19 epidemic. No mean excessive risk from $PM_{2.5}$ was found during the summer of both

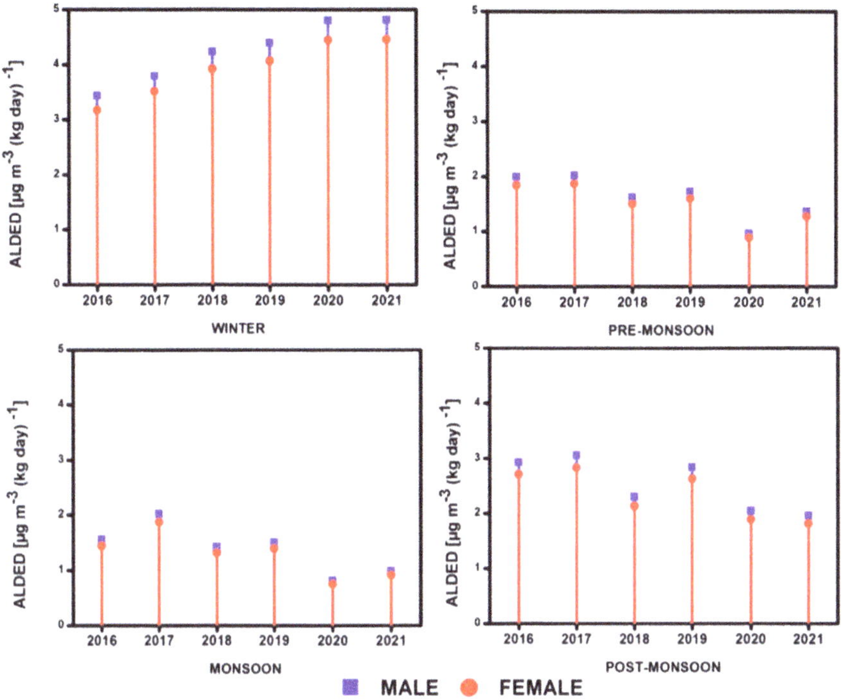

Figure 4: Seasonal ALDED for both males and females.

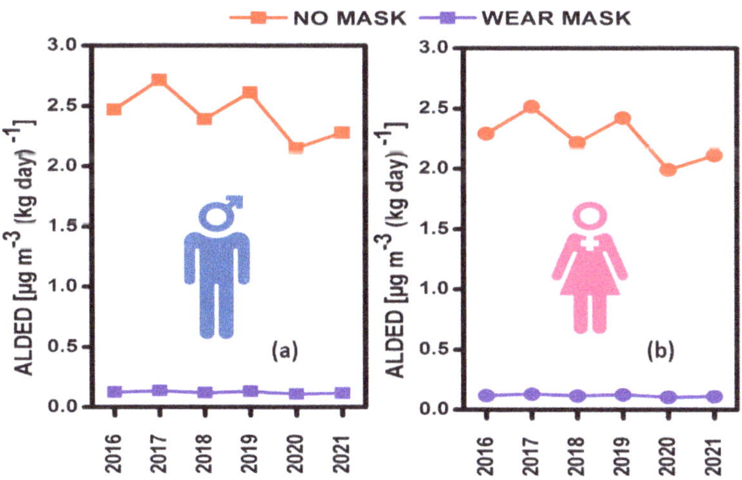

Figure 5: ALDED of: (a) males and (b) females with no mask and after wearing a mask.

2020 and 2021 and the monsoon of 2020, as full or partial restricted activities were imposed during lockdowns. Though the exposure pattern for both males and females followed the same seasonal trend being proportional to the PM concentration level with high exposure dose during winter and lowest during monsoon, however, male exposure doses (d_m) were consistently higher than female exposure doses (d_f) for each exposure condition, despite being of identical magnitude level. Furthermore, when people wear masks, individual health risks are reduced by one or two orders of magnitude. Thus, wearing an ordinary mask also is an efficient way to reduce health hazards by lowering $PM_{2.5}$ concentration intake. The findings of this study's health risk assessment can be used as an educational tool and a policy foundation to enable the adoption of effective preventative efforts to protect public health from environmental health hazards.

References

Craig L, Krewski D, Samet J, Shortreed J, Van Bree L (2007) Strategies for clean air and health. J Toxicol Environ Health A 70(3–4):187–190

Dongarrà G, Manno E, Varrica D, Lombardo M, Vultaggio M (2010) Study on ambient concentrations of PM10, PM10–2.5, PM2.5 and gaseous pollutants. Trace elements and chemical speciation of atmospheric particulates. Atmos Environ 44(39):5244–5257

Flores RM, Özdemir H, Akkoyunlu BO, Ünal A, Tayanç M (2020) Seasonal variation of carbonaceous $PM_{2.5}$ in an Istanbul traffic site. Atmos Pollut Res 11(12):2110–2118

Hansen A, Bi P, Nitschke M, Pisaniello D, Ryan P, Sullivan T, Barnett AG (2012) Particulate air pollution and cardiorespiratory hospital admissions in a temperate Australian city: a case-crossover analysis. Sci Total Environ 416:48–52

Hasan Pasalari, Ataei-Pirkooh A, Aminikhah M, Jonidi Jafari A, Farzadkia M (2019) Assessment of airborne enteric viruses emitted from wastewater treatment plant: atmospheric dispersion model, quantitative microbial risk assessment, disease burden. Environ Pollut 253:464–473

Hu J, Ying Q, Wang Y, Zhang H (2015) Characterizing multi-pollutant air pollution in China: comparison of three air quality indices. Environ Int 84:17–25

Jahne MA, Rogers SW, Holsen TM, Grimberg SJ (2015) Quantitative microbial risk assessment of bioaerosols from a manure application site. Aerobiologia 31:73–87

Kim H, Kim WH, Kim YY, Park HY (2020) Air pollution and central nervous system disease: a review of the impact of fine particulate matter on neurological disorders. Front Public Health 8:575330

Klimont Z, Kupiainen K, Heyes C, Purohit P, Cofala J, Rafaj P, Borken-Kleefeld J, Schöpp W (2017) Global anthropogenic emissions of particulate matter including black carbon. Atmos Chem Phys 17(14):8681–8723

Mahapatra PS, Panda S, Walvekar PP, Kumar R, Das T, Gurjar BR (2014) Seasonal trends, meteorological impacts, and associated health risks with atmospheric concentrations of gaseous pollutants at an Indian coastal city. Environ Sci Pollut Res 21:11418–11432

Mahapatra PS, Sinha PR, Boopathy R, Das T, Mohanty S, Sahu SC, Gurjar BR (2018) Seasonal progression of atmospheric particulate matter over an urban coastal region in Peninsular India: Role of local meteorology and long-range transport. Atmos Res 199:145–158

Nath J, Panda S, Patra SS, Ramasamy B, Das T (2021) Variation of black carbon and particulate matter in Bhubaneswar during the pre-monsoon: possible impact of meteorology and COVID-19 lockdown. Curr Sci 120(2):313

Panda S, Sharma SK, Mahapatra PS, Panda U, Rath S, Mahapatra M, Mandal TK, Das T (2016) Organic and elemental carbon variation in $PM_{2.5}$ over megacity Delhi and Bhubaneswar, a semi-urban coastal site in India. Nat Hazards 80:1709–1728

Panda S, Mallik C, Nath J, Das T, Ramasamy B (2021) A study on variation of atmospheric pollutants over Bhubaneswar during imposition of nationwide lockdown in India for the COVID-19 pandemic. Air Qual Atmos Health 14:97–108

Panda U, Boopathy R, Gadhavi HS, Renuka K, Gunthe SS, Das T (2021) Metals in coarse ambient aerosol as markers for source apportionment and their health risk assessment over an eastern coastal urban atmosphere in India. Environ Monit Assess 193(5):311

Pascal M, Falq G, Wagner V, Chatignoux E, Corso M, Blanchard M, Host S, Pascal L, Larrieu S (2014) Short-term impacts of particulate matter (PM_{10}, $PM_{10–2.5}$, $PM_{2.5}$) on mortality in nine French cities. Atmos Environ 95:175–184

Sharma S, Zhang M, Gao J (2020) Effect of restricted emissions during COVID-19 on air quality in India. Sci Total Environ 728:138878

Shen F, Zhang L, Jiang L, Tang M, Gai X, Chen M, Ge X (2020) Temporal variations of six ambient criteria air pollutants from 2015 to 2018, their spatial distributions, health risks and relationships with socioeconomic factors during 2018 in China. Environ Int 137:105556

Shrivastava D, Nayak B, Kant S, Panda J (2017) A review on health impacts of air pollutants with a special mention to Eastern Parts of India, a growing hub for diseases. Int J Earth Atmos Sci 4:21–33

Tao Y, Mi S, Zhou S, Wang S, Xie X (2014) Air pollution and hospital admissions for respiratory diseases in Lanzhou, China. Environ Pollut 185:196–201

Yan C, Gui ZC, Wu JT (2021) Quantitative microbial risk assessment of bioaerosols in a wastewater treatment plant by using two aeration modes. Environ Sci Pollut Res 28:8140–8150

Zeb N, Khokhar MF, Pozzer A, Khan SA (2019) Exploring the temporal trends and seasonal behaviour of tropospheric trace gases over Pakistan by exploiting satellite observations. Atmos Environ 198:279–290

Effect of Automobile Pollution on Morphological and Biochemical Parameters of Some Tree Plants

Naira Nayab and Md. Anzer Alam

1 Introduction

The typical composition of atmospheric air consists predominantly of oxygen and nitrogen with oxygen around 21% and nitrogen around 78%. Small quantities of water vapour, argon, and carbon dioxide are present, along with trace amounts of other gases like hydrogen and neon. In the contemporary age, air pollution poses a significant threat to the environment. Living organisms grapple with diverse types of pollution, encompassing air, water, soil and noise pollution. The surge in human population, rising automobile numbers, unplanned urbanization, and rapid industrial expansion are the primary contributors to this environmental challenge. The rate of automobiles in India is growing over 6% annually. The vehicle growth rate has reached to approximately 45 million in India. "The vehicle mix is also unique to India in that there is a very high proportion of two-wheelers are 76%" (Kadam & Rajankar, 2011). The expansion of automotive sector plays a pivotal role in driving economic growth, particularly within the Indian automotive industry. "Vehicular pollution contributes to 70% of total air pollution in Delhi, 52% in Mumbai and 30% in Calcutta" (Ravindra et al., 2003; Gokhale and Patil, 2004) "Urban air pollution has become a serious environmental problem to trees and crops" (Chauhan, 2010). "Trees experience the greatest exposure and influenced greatly by pollutant concentration due to their perennial habit" (Chauhan, 2010). It is evident that SO_2 and H_2F_2 are barriers for plant development and can inhibit the yield from various methods. Seeds obtained from regions with higher pollution levels were observed to be of lower quality compared to those from less polluted areas. The decline in the photosynthetic leaf area is attributed to the harmful gases emitted by automobiles. "It is known that SO_2 induced foliar injury; later developing into necrotic lesions hampers growth and decreases the net assimilation rate of plants" (Chand & Singh, 1989). The ongoing

N. Nayab (✉) · Md. A. Alam
Department of Botany, Jai Prakash University, Chapra, Bihar, India
e-mail: nayabnaira@gmail.com

P. K. Padhy et al. (eds.), *Airborne Particulate Matter*,
https://doi.org/10.1007/978-3-031-84408-9_12

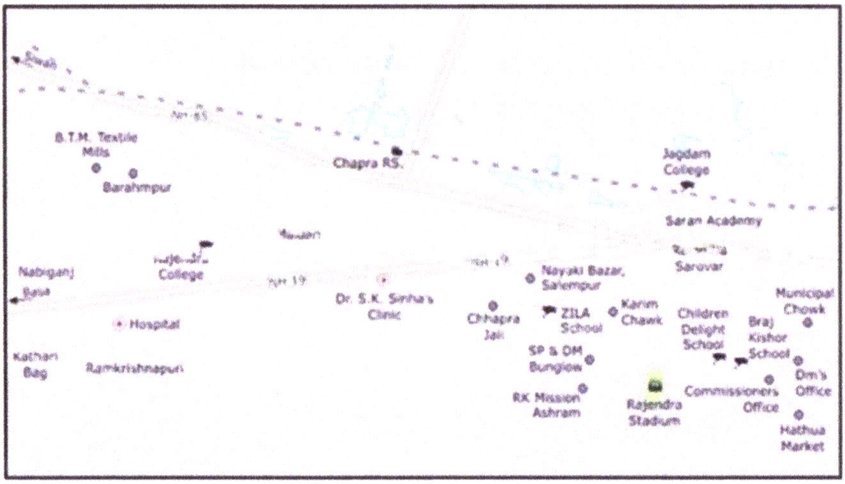

Figure 1: Map of the work area.

discussions have highlighted the detrimental impact of automobile exhaust pollution on plants. With the rapid increase in vehicle population in Chapra, Bihar (India), it has emerged as a significant contributor to air pollution. Hence, this work was conducted to evaluate the air quality, particularly in relation to vehicular emissions, and its impact on the growth of plants.

2 Materials and Methods

2.1 Work Area

Chapra, a small town in western Bihar, India, lies at a latitude of 25.78°N and longitude 84.72°E, with an elevation of 36 m. Positioned along the Ganges and Ghaghara rivers, it faces pollution primarily from automobile emissions. Although not industrial, pollution sources were identified through a survey, considering traffic flow and plant species adaptable to the local environment. Map of the Work Area is shown in Fig. 1. The State Pollution Control Board in Patna, Bihar, monitored air pollutants, providing the data for the study.

2.2 The Control Site-I

University Campus Road and Nehru Chawk Road is taken as Control Site-I or pollution free site area as shown in Fig. 2. The concentration of air pollutants of that area

Figure 2: Control site-I.

was checked through Respirable Dust Sampler (APM 460) operates by drawing air into the suitable reagent daily, throughout each month. This apparatus is utilized for the analysis of Respirable Suspended Particulate Matter and Suspended Particulate Matter.

2.3 *Polluted Site-II*

In Chapra town some roads are found to be always busy and having dense vehicular traffic. The connecting road of Thana Chawk to Kareem Chawk always shows a huge traffic. Similarly connecting road of the Municipal Chawk to the Thana Chawk and the Bhagwan bazaar road to the Thana Chawk Golumber are extensively polluted due to heavy traffic (Fig. 3). Different varieties of plant were grown naturally and some were planted to beautify the place. But due to vehicular emissions plants are constantly exposed and come in contact with the variety of pollutants.

Figure 3: Polluted site-II.

Figure 4: Plant specimens.

2.4 Selection of Plants

Plant samples were collected to evaluate the influence of vehicle discharges on the structural and biochemical characteristics of *Mangifera indica, Ficus religiosa, Polyalthia longifolia* and *Butea monosperma* (Fig. 4).

2.5 Plant Sampling and Analysis

Developed leaves or leaflets of the same age and size were collected from the chosen tree plants at the morning time (5-9 am) and having the same ecological conditions of soil, water and light with extreme caution. Selected plant samples collected from the respective sites were used and wrapped in polythene bags (to minimize evaporation losses). It was brought to the laboratory for further examination. Work was conducted in the three seasons (summer, monsoon and winter) during one year (December 2018 to November 2019). Fresh leaf sample was analyzed for various morphological and biochemical parameters. Air quality parameters were also analyzed during the study period.

2.6 Morphological Parameters

2.6.1 Leaves Colour

Visual observation was done on the randomly collected leaves taken from Site-I and Site-II. Leaves colour were recorded.

2.6.2 Leaf Elongation and Width

The dimensions of the chosen leaves collected from both Site-I and Site-II were gauged using a ruler.

Leaf surface area: Leaf area was determined by assessing the length and width of the leaves and multiplying it with a constant which was derived by determining the leaf area by paper graph method and using relationship

$$L \times B \times C = A$$

$$C = \frac{A}{L \times B}$$

where L is the length of leaf, B is the breadth of leaf, C is the constant of leaf area and A the leaf area.

The seasonal observations were made with 20 leaves from each plant to represent the results.

Petiole length: Petiole length was measured by the help of a ruler.

2.6.3 Biochemical Parameter

Relative water content (%RWC): Three to four leaves were selected and cut into pieces measuring 1 cm^2. The fresh weight of these leaf pieces was measured using a 4″-digit precision scale. Then, all the leave pieces were dipped in water for overnight and weighed again after drying with blotting paper to take turgid weight. Following the measurement of leaf turgid weight, leaf segments were subjected to a 24-hour period in a hot air oven at 80°C, and their weight was subsequently recorded to determine the dry weight. The relative water content was estimated by using the following formula:

$$\text{Relative water content} = (\text{Fresh weight} - \text{Dry weight})/(\text{Turgid weight} - \text{Dry weight}) \times 100.$$

Extraction of protein from sample: The fresh leaf tissue 0.5 g was taken and homogenized in 0.2 M sodium chloride solution containing five milliliter freshly neutralized cysteine-hydrochloric acid. 10 ml of this floating liquid was taken and amalgamated with 10 ml of 10% trichloroacetic acid in a beaker. The solution was then left settled in ice-cold water for 20-30 min. After that the solution was separated and the supernatant was thrown away. Collected precipitate (2 ml) was taken and washed with 70% (v/v) acetone in water. The solution underwent centrifugation once more, and the resulting supernatant was removed. Washing was reiterated until all pigment was removed from the acetone rinse. The washing process was reiterated until all pigment was removed in the acetone rinse. Precipitates were then dissolved in 2 ml of 0.1N NaOH solution and kept in a hot aqua bath for 10 min. Later, it was cooled and centrifuged again. Aliquot from supernatant was used for the estimation of protein. 0.3-0.4 ml of the sample extract was pipette in test tubes for protein estimation.

2.6.4 Ascorbate Content Assessment

The estimation of ascorbic acid content was conducted through the volumetric method. In a 100 ml flask, 5 ml of the working standard solution was combined with 10 ml of 4% oxalic acid and titrated against the dye until a pink colour marked the endpoint (V1 ml). The quantity of dye used in this titration represents the ascorbic acid content. For sample, a two-gram fresh leaf was extracted in a 4% oxalic acid solution, and the volume was adjusted to 100 ml with the same solution. After centrifugation, 5 ml of the supernatant was transferred to a flask, and 10 ml of 4% oxalic acid was added. This solution was then titrated against the dye, and the ascorbic acid content was determined using the specified formula.

Chlorophyll Content Examination: Chlorophyll content was determined using Maclachlan and Zalik (1963) standard methodology.

Investigation of Carotenoid Compounds: The determination of carotenoid content was carried out using the method outlined by Duxbury and Yentsch (1956).

2.6.5 Computation of Air Pollution Tolerance Indices

The methodology provided by Agarwal (1991) was used to calculate the Air Pollution Tolerance Index (APTI) values for tree plants as

$$\text{APTI} = \text{A}\,(\text{P} + \text{T}) + \text{R}/10$$

where A is the content of ascorbic acid, P the pH value, T the total chlorophyll content measured in milligrams per gram of fresh weight (mg/g f wt.) and R is the relative water content expressed as a percentage (%RWC).

Plants were categorized according to the APTI index range, adhering to the tolerance thresholds proposed by Gaikwad et al. (2006). [APTI index scale: 0 to 1 (highly sensitive), 1 to 16 (sensitive), 17 to 29 (intermediate) and 30 to 200 (tolerant).]

Sampling and analyzing air: Air sampling and analysis were conducted at selected locations throughout the designated areas. The sampling and analysis period spanned over one year (December 2018 to November 2019), covering different seasons, namely, summer, monsoon, and winter. The assessment of air quality considered key parameters, like NO_2, SO_2, suspended particulate matter and respirable suspended particulate matter

Table 1: The AQI of Chapra (Bihar India)

Air pollution level in Chapra	*Air Quality Index*	*Main pollutants*
Very unhealthy	251 US AQI	PM 10

2.6.6 Study of Air Quality Measurement

Evaluation of Air Quality Index (AQI) serves as an indicator of the air quality in a particular area. Throughout one year, AQI values for selected locations were computed for each season using the formula.

$$\text{Air Quality Index} = 1/3[(SO_2/SO_2) + (NO_x/NO_x) + (SPM/SPM)] \times 100$$

During that period, the AQI for Chapra town was determined to be significantly unfavourable, as given in Table 1.

2.6.7 Alkalinity Examination

The electrochemical measurement of pH using a pH meter is a valuable method for assessing the acidity and alkalinity of leaf extracts. This test offers insights into the chemical composition of air pollution, providing a comprehensive understanding of its nature:

$$\text{pH} = \log 10\left(H^+\right) \text{ or pH} = \log 10\left(1/H^+\right)$$

2.6.8 Dust Accumulation

In laboratory 11-12 leaves of selected plant species were kept in a beaker and washed with 300 ml distilled water and desiccated at 72°C in a hot air oven. The assessment of dust accumulation involved determining the initial and concluding weights of the beakers, and this was computed using the equation:

$$\text{Weight of dust content}\left(\text{mg/cm}^2\right) = \text{Weight}_2 - \text{Weight}_1 \div \text{Total area of leaf}\left(\text{cm}^2\right)$$

where W_1 is the beaker weight without dust and W_2 the beaker weight with dust.

2.6.9 Arithmetical Examination

Correlation coefficient by Pearson was employed to assess the connection between the investigated air pollutants and meteorological parameters across all sites. The level of variation in the summer, winter and monsoon was assessed using independent groups' t-tests and One-way Analysis of Variance for both control and polluted sites. Statistical analysis was conducted using IBM SPSS Statistics.

3 Results and Discussion

Air pollution impacts on tree plants caused by vehicular exhaust emissions were analyzed by monitoring of air quality of the chosen site areas and the effects of its morphological and biochemical parameters are discussed below. Results of the air quality analysis are given in Tables 2, 3, 4, 5, 6, 7 and 8, which also presents the outcomes of the analysis conducted on diverse external and internal traits of chosen plant species, comparing results between the control and polluted sites.

3.1 Morphological Parameters

3.1.1 Leaf Colour

The colour of leaves in *Mangifera indica* was observed to be dark green in the summer at both polluted and control sites. During monsoon season, it was noted that the foliage appeared light green at the control site, whereas at polluted site, the leaves exhibited a light green hue with noticeable brown spots. In contrast, in summer, the leaves of *Ficus religiosa* were observed to be dark green in colour at both site-I and site-II. During monsoon, the leaves of *Ficus religiosa* were a dark green hue at both site-I and site-II. However, in winter, the leaves of *Ficus religiosa* at the control site retained their dark green colour, whereas those at the polluted site displayed a dark green colour with occasional yellow spots. In summer, the leaves of *Polyalthia longifolia* at control site and polluted site were observed to be dark green. In monsoon, the leaves of *Polyalthia longifolia* displayed a dark green hue at both the control site and the polluted site. However, during winter, the leaves of *Polyalthia longifolia* at control site remained dark green, while at polluted site, they exhibited a dark green colour with occasional brown spots. As for *Butea monosperma*, during summer, the leaves appeared dark green in colour at both the control and polluted sites.

3.1.2 Leaf Length

In summer, leaf length measurement in *Mangifera indica* was found to be 37.133±0.23 cm. The percentage of leaf length measurement was decreased approx. 3.38% at the polluted site compared to the control site. In *Ficus religiosa*, summer season leaf length measurements at control site was observed to be 16.033 ± 0.50

Table 2: Study of chlorophyll pigments

Variables	*Ficus religiosa*		*Mangifera indica*		*Polyalthia longifolia*		*Butea monosperma*	
	Reference site	*Pollution-affected area*	*Reference site*	*Pollution-affected area*	*Reference site*	*Pollution-affected area*	*Reference site*	*Pollution-affected area*
Chlorophyll-*a* levels in mg/g	1.43±0.08	0.81±0.06	2.71±0.07	1.99±0.05	1.33±0.17	0.96±0.44	1.64±0.11	1.04±0.11
Chlorophyll-*b* levels in mg/g	1.13±0.05	0.61±0.05	2.03±0.09	1.63±0.12	0.84±0.12	0.59±0.06	1.01±0.15	0.76±0.14
Total chlorophyll content (mg/g)	2.56±0.48	1.42±0.09	4.74±1.22	3.62±0.58	2.17±0.22	1.55±0.28	2.65±0.14	1.81±0.12

Table 3: Analysis of carotenoid

Plant species	*Control site (mg/gm)*	*Polluted site (mg/gm)*	*Decrement in carotenoid (%)*
Ficus religiosa	1.49±0.14	1.05±0.08	40.80
Mangifera indica	2.65±0.15	2.20±0.18	20.45
Polyalthia longifoli	1.41±0.15	0.99±0.18	30
Butea monosperma Regia	1.30±0.08	0.90±0.06	30.78

Table 4: Analysis of APTI

APTI	*Control site*	*Polluted site*
Ficus religiosa	8.40±0.18	7.25±0.19
Mangifera indica	9.16±0.12	6.72±0.15
Polyalthia longifolia	7.02±0.15	5.58±0.20
Butea monosperma	8.49±0.18	6.60±0.29

Table 5: Analysis of ascorbic acid

Plant species	*Control site (mg/gm)*	*Polluted site (mg/gm)*	*Decrement in ascorbic acid (%)*
Ficus religiosa	1.55±0.15	0.85±0.09 mg per 50 g	45.68
Mangifera indica	2.18±0.12	1.58±0.15 mg per 50 g	18.25
Polyalthia longifolia	1.35±0.08	1.05±0.07 mg per 100 g	22.39
Butea monosperma	3.46±0.09	2.06 mg ± 0.09 mg per 100 g	28.30

Relative Moisture Content, pH and APTI

Table 6: Analysis of pH

pH (plant species)	*Control site*	*Polluted site*	*Total change towards acidic side (%)*
Ficus religiosa	8.68±0.15	6.45±0.25	15.29
Mangifera indica	6.95±0.23	6.01±0.23	14.29
Polyalthia longifolia	5.48±0.14	5.75±0.25	12.45
Butea monosperma	6.95±0.38	5.65±0.28	18.50

Table 7: Analysis of dust content

Dust content (plant species)	*Control site*	*Polluted site*
Ficus religiosa	0.12±0.04 mg per leaf	0.45±0.03 mg per leaf
Mangifera indica	0.08±0.02 mg per leaf	0.55±0.03 mg per leaf
Polyalthia longifolia	0.09±0.03 mg per leaf	0.44±0.06 mg per leaf
Butea monosperma	0.09±0.05 mg per leaf	0.40±0.08 mg per leaf

Table 8: Airborne pollutants detected at both control and contaminated locations

Site	*NO_2 μgm-3*			*SO_2 μgm-3*			*SPM (μgm-3)*			*RSPM(μgm-3)*			*AQI*		
	W	*M*	*S*	*W*	*M*	*S*	*W*	*M*	*S*	*W*	*M*	*S*	*W*	*M*	*S*
Polluted Sites	26.68	27.92	27.92	25.14	26.61	27..11	228.04	190.47	280,24	130.40	90.70	130.70	110,13	90.40	120.80
Control Sites	1.70	1.59	1.90	2.30	1.98	2.18	90.20	80.15	70.20	30.40	30.85	30.45	30.40	20.80	28.15

cm while at polluted site, its value was found to be 15.567±0.61 cm. The percentage amount was observed to be decreased (approx. 2.91%). In contrast to the control site, the polluted site showed a reduction in measurements. In monsoon, the measurements of leaf length in *Ficus religiosa* was observed to be 18.65±0.40 cm at control site while at polluted site, it was seen to be 15.2±0.69 cm; the percent decrease at polluted site in comparison to control site was observed to be 4.25%. During summer the measurements for leaf length in *Polyalthia longifolia* was observed to be 21.36 ±1.65 cm at control site while at polluted site, it was found to be 19.67±0.64 cm; the percent decreased at polluted site in comparison to control.

3.1.3 Leaf Area

In summer, *Ficus religiosa* showed a leaf area of 145.68±4.75 cm^2 at the control site, while at the polluted site it measured 142±5 cm^2, indicating a percentage decrease of 7.08% compared to the control site. During monsoon, the control site recorded a leaf area of 144 ±10.46 cm^2, whereas the polluted site showed a value of 145.33±1.55 cm^2, resulting in a percentage decrease of 7.33%. In winter, *Ficus religiosa* displayed a leaf area of 128±6.98 cm^2 at the control site, while at the polluted site, it measured 108 ±9 cm^2, indicating a percentage decrease of 18.50% compared to the control site. For *Polyalthia longifolia*, the leaf area was 44±4.48 cm^2 at the reference site, and at the polluted site, it was 44±4.58 cm^2, indicating a percentage decrease of 5.45% compared to the reference site. In *Butea monosperma*, the leaf area was 97 ±4.58 cm^2 at the control site, while at the polluted site, it measured 86.33 ±3.21 cm^2, resulting in a percentage decrease of 12.00% compared to the control site.

4 Conclusion

Air pollutants adversely affected the leaves of the plants, resulting in damages of chlorophyll, carotenoids, ascorbic acid contents and thus finally decreased the plant productivity. The results shows that the toxicants and particulate pollutants give deleterious impacts on tree plants. Alterations in the levels of air pollution at specific locations are directly linked to changes in photosynthetic pigments. The research elucidates that emissions from automobiles contribute to the degradation of air quality, impacting both ambient air conditions and the pigmentation of trees. Furthermore, these findings underscore the detrimental consequences of air pollution on human health.

References

Agarwal M, Singh SK, Singh J, Rao DN (1991) Biomonitoring of air pollution around urban and industrial sites. J Environ Biol 211–212

Chand S, Singh V (1989) Sensitivity of Triticale hexaploide Cv. Panda-6 to sulphur dioxide. Proc Plant Sci 99:271–278

Chauhan A (2010) Tree as bioindicator of automobile pollution in Dehradun City: a case study. N Y Sci J 3(6):88–95

Duxbury AC, Yentsch CS (1956) Plankton pigment nomographs. J Mar Res 15, 1, 866

Gaikwad A, Maitra A, Zhang P, Ingram M, Mercado DL, Woit WD (2006) Using system disturbance measurement data to develop improved load models. IEEE PES Power Syst Conf Expo 1978–1985

Gokhale SB, Patil RS (2004) Size distribution of aerosols (PM_{10}) and lead (Pb) near traffic intersections in Mumbai (India). Environ Monit Assess 95(1):311–324

Kadam ST, Rajankar SO (2011) Automatic start stop system for two wheelers to improve fuel economy. Glob J Mech Eng Comput Sci 1(3):22–24

Maclachlan S, Zalik S (1963) Plastid structure, chlorophyll concentration, and free amino acid composition of a chlorophyll mutant of barley. Can J Bot 41(7):1053–1062

Ravindra K, Mor S, Ameena, Kamyotra JS, Kaushik CP (2003) Variation in spatial pattern of criteria air pollutants before and during initial rain of monsoon. Environ Monit Assess 87:145–153

Effect of Particulate Matter Pollution on Rice Crop and Greenhouse Gas Emissions from Rice Fields

Debasish Rout, Manisha Rout, Pritam Ghosh, and S. N. Jena

1 Introduction

Oryza sativa (rice) is a staple food for over half of the world's population (Bhar et al., 2022). Its cultivation takes place in diverse agro-ecosystems, and various factors impact its growth and yield. One emerging concern is the influence of particulate matter on rice crops and the subsequent implications for greenhouse gas emissions.

2 Particulate Matter and its Sources

Particulate matter (PM) consists of fine particles suspended in the air, categorized as $PM_{2.5}$ (particles with a diameter of 2.5 micrometers or smaller) and PM_{10} (particles with a diameter of 10 micrometers or smaller). Natural sources (e.g., dust, pollen) and anthropogenic sources (e.g., industrial emissions, vehicular exhaust) contribute to PM pollution. The deposition of PM on rice plants can directly affect their physiology and yield (Malav et al., 2022; Falco et al., 2020).

D. Rout
Synergy Institute of Engineering and Technology, Dhenkanal, Odisha, India

M. Rout (✉) · P. Ghosh
Institute of Agriculture, Visva-Bharati University, Bolpur, West Bengal, India
e-mail: manisha.rout55@gmail.com

S. N. Jena
Department of Agronomy, College of Agriculture, University of Agriculture and Technology, Bhubaneswar, Odisha, India

P. K. Padhy et al. (eds.), *Airborne Particulate Matter*,
https://doi.org/10.1007/978-3-031-84408-9_13

3 Effect on Rice Crop

The PM deposition on rice leaves can hinder photosynthesis, limit light penetration, and obstruct stomatal function, leading to reduced crop growth and productivity. Moreover, PM can alter the physicochemical properties of the soil, affecting nutrient availability and water retention, both critical for rice growth.

Numerous studies have investigated the impact of PM deposition on rice plants and their subsequent effects on plant physiology and yield. Some common findings and trends from the literature are (Malav et al., 2022):

- **Direct Physical Effects:** The PM deposition on rice leaves can physically block stomata, reducing gas exchange and photosynthesis. This can lead to decreased chlorophyll content and a decline in photosynthetic efficiency (Malav et al., 2022; Falco et al., 2020., Mathur et al., 2023).
- **Nutrient Imbalance:** The PM particles can contain heavy metals and other pollutants that can be absorbed by rice plants. These pollutants can disrupt nutrient uptake and lead to nutrient imbalances, affecting plant growth and yield (Mathur et al., 2023; Tao et al., 2023) (Fig. 1).

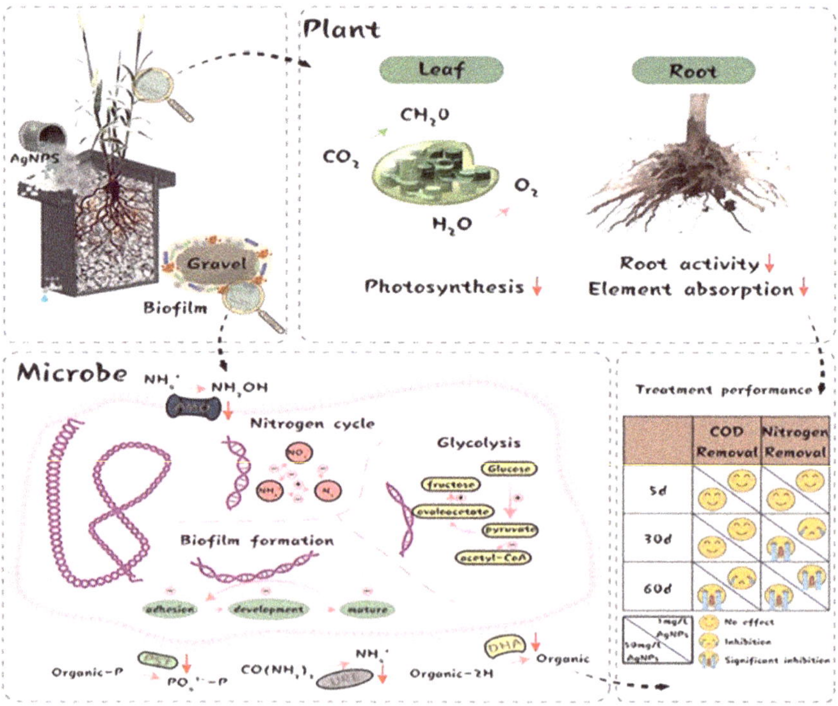

Figure 1: Effect of PM on nutrient absorption by plant roots (Liu et al., 2023)

- **Oxidative Stress:** The PM deposition can induce oxidative stress in rice plants due to the production of reactive oxygen species (ROS). This stress can damage cellular structures and impair various physiological processes, potentially leading to reduced yield (Liu et al., 2023; Elina et al., 2021).
- **Leaf Morphology:** High levels of PM deposition can alter leaf morphology, including size and shape. Changes in leaf morphology can affect light interception, which in-turn influences photosynthesis and yield (Elina et al., 2021; Xiaosan et al., 2019).
- **Disease Susceptibility:** The PM deposition can create a microenvironment conducive to the growth of pathogens. Weakened plant defenses due to PM-induced stress can increase the susceptibility of rice plants to diseases (Quanhui et al., 2022; Xiaosan et al., 2019).
- **Yield Reduction:** Accumulated effects of PM deposition, such as reduced photosynthesis, nutrient imbalance and oxidative stress, can contribute to overall yield reduction in rice crops (Bhar et al., 2022; Malav et al., 2022; Mathur et al., 2023).
- **Genotypic Variability:** Different rice cultivars show varying levels of susceptibility to PM deposition. Some cultivars exhibit more resilience to the negative effects of PM, while others are more susceptible (Bhar et al., 2022; Liu et al., 2023; Elina et al., 2021).
- **Mitigation Strategies:** Some studies have explored the use of physical barriers, such as nets or screens, to reduce direct PM deposition on rice leaves. Additionally, adopting proper agricultural practices that minimize PM emissions and exposure can mitigate negative effects on rice physiology and yield (Xiaosan et al., 2019).
- **Interaction with Other Stressors:** The effects of PM deposition on rice plants can interact with other environmental stressors, such as drought, temperature extremes and chemical pollutants. These interactions can lead to compounded effects on plant health and yield (Sillmann et al., 2021).

Overall, the literature indicates that PM deposition on rice plants can have multifaceted and complex effects on their physiology and yield. The extent of these effects depends on factors such as PM concentration, particle size, composition, rice cultivar and environmental conditions. Understanding these interactions is essential for developing effective strategies to mitigate the negative impacts of PM deposition on rice crops and enhance agricultural sustainability.

4 Mechanisms of Impact

The impact of PM on rice crops is multifaceted. It can cause oxidative stress due to the generation of reactive oxygen species, leading to cellular damage. PM-induced stress responses in rice include changes in gene expression, enzymatic activity, and hormonal regulation, all of which collectively influence growth and yield (Bhar et al., 2022; Xiaosan et al., 2019; Sillmann et al., 2021).

5 Greenhouse Gas Emissions

Rice fields with inundated irrigation water are unique ecosystems that emit significant amounts of greenhouse gases, including methane (CH_4) and nitrous oxide (N_2O). The presence of PM on rice plants and in paddies can influence these emissions by affecting soil aeration, microbial activities, and nutrient cycling. PM deposition can both enhance and mitigate GHG emissions, depending on the specific conditions and interactions (Bhar et al., 2022; Malav et al., 2022; Mathur et al., 2023; Quanhui et al., 2022).

6 Mitigation Strategies

To mitigate the negative effects of PM on rice crops and its impact on GHG emissions, several strategies can be employed. These include improving air quality through emission control measures, adopting sustainable agricultural practices that reduce soil disturban (Mohammadi et al., 2020; Singh et al., 2021).

6.1 Mitigation Strategies to Alleviate PM Effects on Rice Crops

- Following are the mitigation strategies to alleviate PM effects on rice crops:
- *Cultural Practices:* Adoption of sustainable cultivation practices such as agroforestry, cover cropping and crop rotation can reduce PM deposition on rice leaves, enhancing plant health and productivity.
- *Biological Solutions:* Incorporation of beneficial microorganisms, such as mycorrhizal fungi and endophytes, can enhance plant resistance to PM-induced stress.
- *Physical Barriers:* Implementation of screens, nets or agrotextiles can reduce direct PM deposition on rice plants, preserving leaf health and photosynthetic efficiency.
- *Air Quality Improvement:* Enforcing stricter emission controls on industrial activities and vehicular emissions can lower ambient PM levels, reducing the overall PM load on rice crops.

6.2 Mitigation Strategies to Influence GHG Emissions (Miller et al., 2021)

The mitigation strategies to influence GHG emissions are:

- *Alternate Wetting and Drying (AWD):* The AWD water management reduces methane emissions by intermittently allowing the soil to dry, suppressing methanogenic activity while maintaining rice productivity.
- *Innovative Rice Cultivation:* Practices like system of rice intensification (SRI), aerobic rice cultivation or non-flooded direct seeding can lower methane emissions, as these systems create conditions unfavourable for methanogenesis.
- *Nitrogen Management:* Improved nitrogen use efficiency through precise fertilizer application and organic matter incorporation can minimize nitrous oxide emissions from rice fields.

6.3 Synergistic Approaches

Following are the synergistic approaches:

- *Integrated Pest Management (IPM):* Combining IPM practices with PM reduction measures can enhance rice plant health, reducing the need for pesticide applications that contribute to GHG emissions.
- *Cover Crops and Residue Management:* Integrating cover crops and residue management practices can decrease PM deposition, improve soil health and reduce the need for tillage, thus mitigating GHG emissions.

6.4 Climate-Smart Agriculture

Recently climate resilient crops and climate smart agricultural practices are gaining momentum which can significantly reduce GHG emissions from agricultural crops and practices as mentioned below:

- *Precision Agriculture:* Utilizing technology for site-specific nutrient application reduces PM deposition and mitigates nutrient-related GHG emissions, enhancing both air quality and soil health.
- *Ecosystem-based Adaptation:* Incorporating natural ecosystems like wetlands or buffer zones around rice fields can mitigate PM effects, while also serving as carbon sinks.

7 Future Research Directions

Further research is needed to comprehensively understand the intricate interactions between PM pollution, rice crops, and GHG emissions. Long-term studies evaluating the effectiveness of different mitigation strategies under various environmental conditions are necessary to develop targeted interventions.

8 Conclusion

Particulate matter pollution poses significant challenges to rice crop health and greenhouse gas emissions from rice fields. Addressing this issue requires a multidisciplinary approach involving agronomists, environmental scientists, policy makers and stakeholders. Mitigating the impact of particulate matter on rice crops is important not only for ensuring global food security but also for promoting sustainable agricultural practices and reducing greenhouse gas emissions.

References

Bhar S, Bose T, Dutta A et al (2022) A perspective on the benefits of consumption of parboiled rice over brown rice for glycaemic control. Eur J Nutr 61(2):615

Falco WF, Scherer MD, Oliveira SL, Wender H, Colbeck I, Lawson T, Caires ARL (2020) Phytotoxicity of silver nanoparticles on *Vicia faba*: Evaluation of particle size effects on photosynthetic performance and leaf gas exchange. Sci Total Environ 701:134816 ISSN 0048-9697

Liu T, Guo F, Chen M, Zhao S, Yang X, He Q (2023) Silver nanoparticles disturb treatment performance in constructed wetlands: responses of biofilm and hydrophyte. J Clean Prod 385:135751 ISSN 0959-6526

Luo X, Bing H, Luo Z, Wang Y, Jin L (2019) Impacts of atmospheric particulate matter pollution on environmental biogeochemistry of trace metals in soil-plant system: a review. Environ Pollut 255(1):113138. https://doi.org/10.1016/j.envpol.2019.113138. ISSN 0269-7491

Ma Q, Li Y, Zhu Y, Liu X, Yu H, Li L, Qi M et al (2022) Precipitation variations, rather than N deposition, determine plant ecophysiological traits in a desert steppe in Northern China. Ecol Indic 141:109144 ISSN 1470-160X

Malav LC, Kumar S, Islam S et al (2022) Assessing the environmental impact of air pollution on crops by monitoring air pollution tolerance index (APTI) and anticipated performance index (API). Environ Sci Pollut Res 29:50427–50442. https://doi.org/10.1007/s11356-022-19505-3

Mathur P, Chakraborty R, Aftab T, Roy S (2023) Engineered nanoparticles in plant growth: Phytotoxicity concerns and the strategies for their attenuation. Plant Physiol Biochem 199:107721 ISSN 0981-9428

Miller SA, Habert G, Myers RJ, Harvey JT (2021) Achieving net zero greenhouse gas emissions in the cement industry via value chain mitigation strategies. One Earth 4(10):1398–1411

Mohammadi A, Khoshnevisan B, Venkatesh G, Eskandari S (2020) A critical review on advancement and challenges of biochar application in paddy fields: environmental and life cycle cost analysis. Processes 8(10):1275

Oksanen E, Kontunen-Soppela S (2021) Plants have different strategies to fields: environmental and life cycle cost analysis. Processes 8(10):1275

Sillmann J, Aunan K, Emberson L, Büker P, Van Oort B, O'Neill C, Otero N, Pandey D, Brisebois A (2021) Combined impacts of climate and air pollution on human health and agricultural productivity. Environ Res Lett 16(9):093004

Singh G, Gupta MK, Chaurasiya S, Sharma VS, Pimenov DY (2021) Rice straw burning: a review on its global prevalence and the sustainable alternatives for its effective mitigation. Environ Sci Pollut Res 28(25):32125–32155

Assessment of Heavy Metal Bioremediation Potential of Plants and Microbes: A Review

Shubhasree Singha and Pratap Kumar Padhy

1 Introduction

In this era of modernisation, rapid industrialisation, urbanisation and population explosion have caused the introduction of various contaminants into all spheres of the environment; this has significantly altered the physical and chemical characteristics of land, water and air, which adversely impacted the biotic community. Heavy metals (HMs) are one pollutant class that severely threatens the environment and food security. Heavy metal pollution in soil refers to the build-up of toxic heavy metals in soil from various sources, such as industrial activities, mining, and agricultural practices. Heavy metals can persist in soil for a long time and threaten human health and the environment.

Bioremediation is one strategy used to address soil's heavy metal pollution. It entails removing or degrading heavy metals from the earth and utilising living things like plants and microorganisms. Phytoremediation is bioremediation that eliminates or neutralises pollutants from soil and water using plants and the microorganisms that live with them. Plants take pollutants through their roots or leaves, either stored in plant tissue or broken into less toxic forms. Phytoremediation is a less expensive and more ecologically acceptable alternative to traditional remediation methods. An affordable and environmental friendly substitute for conventional remediation techniques is phytoremediation. It is capable of eliminating a wide range of pollutants, such as organic pollutants, radioactive elements and heavy metals. Microbes are used in microbial remediation to break or change heavy metals into less harmful forms.

Various techniques can be employed to evaluate the capacity of plants and bacteria for bioremediation of heavy metal-contaminated soil, including phytotoxicity tests, microbial toxicity tests, biodegradation assays, and bioaccumulation investigations.

S. Singha · P. K. Padhy (✉)
Department of Environmental Studies, Siksha Bhavana (Institute of Science), Visva-BharatiSantiniketan, Birbhum, West Bengal, India
e-mail: pkpadhy@visva-bharati.ac.in

P. K. Padhy et al. (eds.), *Airborne Particulate Matter*,
https://doi.org/10.1007/978-3-031-84408-9_14

Determining a plant or microbe's potential for bioremediation entails analysing its growth and survival in polluted surroundings and its ability to break contaminants. Laboratory experiments, field research and modelling simulations can all be used to achieve this. Field studies can also be carried out to assess the efficiency of bioremediation under actual environmental circumstances.

In conclusion, evaluating a plant or microbe's capacity for bioremediation is essential to figuring out how efficient it will be in cleaning up polluted areas. It aids in optimising the application of various bioremediation techniques and offers valuable information on their applicability. It is important to note that bioremediation is not a standalone solution for heavy metal pollution. It should be combined with other remediation technologies, such as physical and chemical treatments, to achieve the best results. This article reviewes various studies on phytoremediation conducted in recent years.

2 Heavy Metal

The HMs are metals that have a density more than or equal to 5 gm/cm^3 and atomic numbers above 20, as per major consensus. The HMs are classified broadly into two categories: essential and non-essential. Essential HMs are those required up to a specific concentration by living organisms for fundamental processes like metabolism, growth, and development. HMs like copper (Cu), iron (Fe), manganese (Mn), cobalt (Co), zinc (Zn) and nickel (Ni) fall into the essential category for plants and animals. However, these can also cause metal stress and toxicity in higher concentrations. Many of the non-essential HMs like arsenic (As), cadmium (Cd), chromium (Cr), lead (Pb), mercury (Hg), and selenium (Se) are seen as cytotoxic, mutagenic and carcinogenic, even in lower concentrations. HMs are not biodegradable and can remain in the environment for extended periods, easily pass through various pathways, and even transform into more toxic compounds through multiple processes.

The HMs have both natural and anthropogenic sources, but in recent times, rapidly increasing human activity has disrupted natural geochemical cycles and biochemical balance. Industrial effluents, mining, smelting, biomass burning, fuel extraction and refining, military exercise, domestic pollutants and agricultural chemicals are the primary anthropogenic sources resulting in the introduction of HMs to land, water and air. Soil and water play a significant role as sinks for these HMs (Fayiga & Saha, 2016; Zhao et al., 2016). Airborne HMs are present as aerosols in the atmosphere, some of which can directly enter living organisms through different channels and cause drastic health implications. Most airborne HMs gradually settle down, wash out in precipitation, and come down to water and soil.

Moreover, HMs in water come into the soil through irrigation and then leach into the soil. These HMs gradually accumulate in living organisms and enter the food web, giving rise to biomagnification and creating severe consequences. This way, HMs can have dangerous repercussions for health and the environment.

3 Remediation Techniques

With increased awareness about HMs, various techniques have been implemented to abate environmental pollution. These techniques are broadly divided into physical, chemical and biological methods (Lim et al., 2014). Physical and chemical remediations have severe disadvantages over bioremediations, as these are very costly and energy-intensive, produce secondary pollution, alter soil and water properties, and disrupt microbial communities (Ali et al., 2013). In contrast to the earlier two, bioremediation techniques are safer, innovative, low-risk, and economically feasible alternatives preferred to control HM pollution.

3.1 Bioremediation

Bioremediation includes numerous techniques that use living or dead organisms like plants, animals and microbes to reduce environmental pollution. Among all techniques, phytoremediation has gained immense attention lately (Yang et al., 2020; Liu et al., 2020a, b); translocation, accumulation, transport, transformation, and volatilisation processes are included (Shah & Davey, 2020; Wang et al., 2012; Wu et al., 2010).

3.2 Processes of Phytoremediation

Phytoremediation of metals carry different physiological processes; based on this, it is categorised into phytoextraction, phytostabilization, phytofiltration, phytotransformation and phytovolatilisation (Fig. 1). By using a technique called phytoextraction, plants may collect and gather heavy metals in their shoots without changing the characteristics of the soil. Thus, HMs can be eliminated from contaminated sediments, water or soil by harvesting these plants (Liu et al., 2020a, b; Ali et al., 2013). Through a process called phytostabilisation, plants can convert HMs from a toxic to a non-toxic or less hazardous form (Nwoko, 2010; Li et al., 2019). The phytofiltration technique includes chelation, redox reaction, adsorption precipitation and absorption in the rhizosphere (Wu et al., 2010; Ghosh & Singh, 2005; Wetle et al., 2020). Phytotransformation or phytodegradation, is the plant's breakdown of pollutants ingested by its metabolic processes or broken by the medium's root exudates (like enzymes). Mainly, it eliminates organic contaminants with moderate hydrophobicity (Al-Baldawi et al., 2018). Phytovolatilisation is a process, in which pollutants are taken up by plant roots, converted into a gas by the plant metabolism, and then released into the atmosphere. It is the most controversial phytoremediation technique due to its apparent disadvantages and limitations as it reintroduces HMs to the environment in gaseous form (Padmavathiamma & Li, 2007; Sarma, 2011). Hg, As and

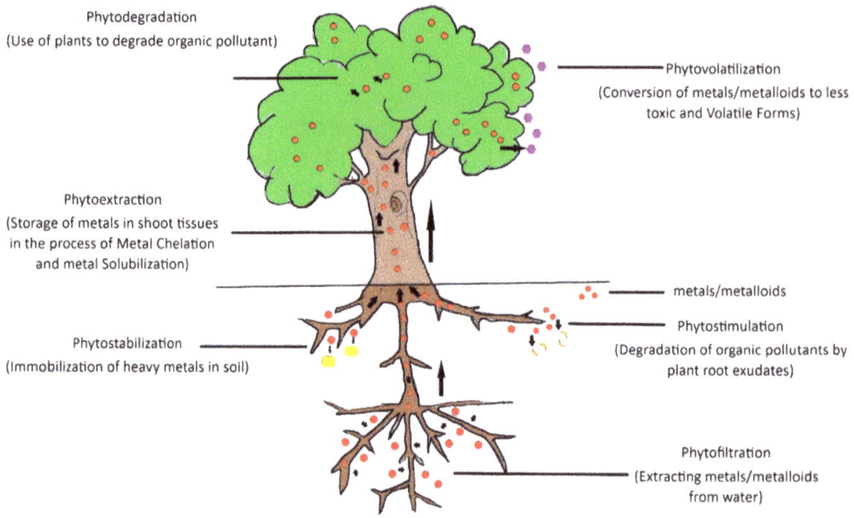

Figure 1: Phytoremediation Processes.

Se are among the HMs that are primarily removed by this method because of their high volatility and low toxicity (Wang et al., 2012; Sharma et al., 2015; Guarino et al., 2020).

3.3 *Factors Affecting Phytoremediation*

According to Wei et al. (2008), key to the success of phytoremediation technology is choosing right species. The plant species ought to possess a strong ability to absorb and accumulate heavy metals (HMs), a high biomass, and a rapid rate of development. These qualities are determined by various factors, including soil's properties, rhizospheric microbial population, climate, and the HMs' bioavailability (Odokuma & Akponah, 2010; Meharg & Cairney, 2000). The bioavailability of HMs can be impacted by oxygen, moisture, pH, organic matter, and other soil variables. This can directly impact the effectiveness of phytoextraction (Wei et al., 2008). Metals in soils are classified as accessible, unavailable or exchangeable based on their absorbability (Wei et al., 2008; Ashraf et al., 2019). Various studies have shown that pH primarily adjusts the dissolution of HMs in soil (Javed et al., 2019; Ashraf et al., 2019), whereas high pH (alkalinity) reduces the metals' bioavailability by lowering their dissolution (Zeng et al., 2017). Various parameters that influence or limit phytoremediation efficiency shown in recent studies are given in Table 1.

Impact of physical factors like pH, metal concentration, and planting density and chemical factors like application of biochar, digestate, urea, and multi-walled carbon nanotubes and biological factors like specific genotype of plants and application of

Table 1: Physical factors affecting Phytoremediation

Target	*Plant*	*Influence/enhancement factor*		*Result*	*References*
		Category	*Detail*		
Soil	*Helianthus tuberosus*	pH Metal concentration	4.00-6.00 Slight/ medium/high	At pH 5.00 and pH 6.00, *Helianthus tuberosus* was shown to develop enough at modest metal concentrations. The amount of metal in the soil at pH 5.00 rose in proportion to the metal concentrations in the shoots.	Willseher et al. (2017)
Soil	*Vicia Faba* L.	Genotype	Ten genotypes	Most suitable genotype for phytoremediation of soil with mild to moderate co-contamination of lead and copper was shown to be LXYC.	Tang et al. (2019)
Soil	*Festuca arundinacea*	Planting density	10, 15, 20, 25 and 30 g seeds m^{-2}	At D20, the biomass and Cd accumulation peaked (13.30 g Cd' m-2).	Qin et al. (2021)
Soil	*Solanum nigrurn* L.	Multi-walled carbon nanotubes	100, 500 and 1000 mg kg^{-1}	Carbon nanotubes with many walls have the potential to enhance plant biomass by 5.23-27.07%.	Chen et al. (2021)
Soil	*Bidens pilosa* L.	Biochar	100 and 200 mg kg^{-1}	Treatments with 100 and 200 mg kg^{1} of biochar enhanced the accumulation of Cd by 16.44 and 39.37%, respectively.	Manori et al. (2021)
Soil	*Pennisetum hybridum*	Digestate	0%, 25%, 50%, 75% and 100%	Digestate addition increases Cd contents in the stems (7.94-42.39%), leaves (12.53-74.11%), and roots (18.59-57.94%).	He et al. (2021)

(continued)

Table 1: (continued)

Target	*Plant*	*Influence/enhancement factor*		*Result*	*References*
		Category	*Detail*		
Soil	Willow	Urea	0 and 0.40 g kg^{-1}	Following Cd and urea treatment, the Cd concentrations in the roots, xylem, bark and leaves were 8.30, 8.15, 26.79 and 33.04 mg/kg, respectively.	Li et al. (2021).
Water	*Lemna valdiviana*	pH $P\text{-}PO_4$ $N\text{-}NO_3$	3.94-9.02 0-0.14 $mmol.L^{-1}$ 0.09-13.71 $mmol.\ L^{-1}$	*Lemna valdiviana* collects As (1190 mg kg^1), $P\text{-}PO_4$ (0.05 $mm^{1\ L1}$), and $N\text{-}NO_3$ (7.90 $mmolL^1$) at much higher concentrations when the pH is between 6.30 and 7.00.	Souza et al. (2019)
Water	*Utva ohnoi*	Temperature Salinity	18, 25, 28 S15, S35	*Ulva ohnoi* continued to develop positively between 180°C, S15 and 180-250°C, S35. The highest concentration factor reached 81.30% when 0.63 μgL^{-1} of Cd was added to 180°C and S15.	Bastos et al. (2019)

genetically engineered yeast on phytoremediation efficiency of different plants on soil has been shown below. Following are some reports from recent studies showing how alterations in physical parameters like temperature, water salinity, pH and phosphate and nitrate concentrations might affect phytoremediation.

3.4 Phytoremediation by Various Species

We select species with rapid growth rates, higher biomass, resistance to metal stress and high HM accumulation potential for phytoremediation and also confirm that those selected species do not disrupt the balance of local flora. We also have to consider lower land use, the shorter life-cycle of species, the lower cost of plantation, and further implementation for successful field application of phytoremediation. Results of recent research on plant phytoremediation efficiency are given in Table 2.

Table 2: HM bioremediation potential of different plant species

Plants	*Metal*	*Concent-ration (mg/kg)*	*Remediation time (days)*	*Metals in plants (mg/kg)*			*Reference*
				Shoot		*Root*	
				Stem	*Leaf*		
Centella asiatica L.	Cd	18.4	90	77.7	95.5	38.9	Li et al. (2018).
Napier grass	Cd	50.4	180	38.1			Wiangkham and Prapagdee (2018)
Tagetes erecta L.	Cd	160	180	318		479	Minisha et al. (2021)
Sedum plumbizincicola	Cd	1.3	100	522			Hou et al. (2019)
Khaya ivorensis	Cu	600	90	52		329	Covre et al. (2020)
Cedrela fissilis	Cu	600	90	26		216	Covre et al. (2020)
Leersia hexandra swartz	Cu	480	60	650.0-700.0		5501.7	Lin et al. (2019)
Corontpus didymus L.	Zn	500	45	1845		1848	Sidhu et al. (2020)
Bidens pilosa L.	Ni	24.5	90	27.8	52.6	180	Liu et al. (2019)
Arundinaria argenteostriata	Pb	1500	90	845.7	540.7	1117.7	Jiang et al. (2019)
Sesbania rostrata	U	80	60	33.2		32.8	Ren et al. (2019)
Pinus thunbergii	Cr(VI)	800.O	150	512.4		2317.8	Shi et al. (2019)
Melilotus officinalis	Cu	3480.3		1043.3		704.7	Ghazaryan et al. (2021)
	Mo	1526.5		212		259	
Pteris cretica	As	200	30	4875			Eze & Harvey (2018)
	Pb	160		10.6		54.8	
	Zn	160.O		174		690	
Grapevine	Cu	1184.8	730	910.2		242	Mirzaei et al. (2021)
	Pb	865.5		152.9		63.5	
	Cd	803.2		322.9		239.5	
	Cr	623.O		25.3		17.2	
	Zn	277.9		160.3		188.8	

The result of research conducted in recent years on phytoremediation for different plant species on their efficiency in HMs removal from soil and bioaccumulation in their plant parts (root, stem, leaf) for different time periods is given below.

3.5 Microbial-assisted Phytoremediation

The rhizosphere bacteria are essential to plants' growth and well-being (Ali et al., 2017; Mendoza-Hernández et al., 2019). Some discourage the concentration of HMs and encourage the growth of plants (Wiangkham & Prapagdee, 2018). Some of their metabolites can make metals in the soil more bioavailable (Li et al., 2020). Numerous phytoremediation procedures, including phytoextraction, phyto-stabilisation, and phyto-transformation, involved the use of microorganisms.

A study by Li et al. (2019) on the phytoremediation of Pb application of *Trichoderma asperellum* fungus in the *Suaeda salsa* plant's rhizosphere resulted in a 9-23% and 5-13% increase in plant height and fresh weight, respectively. *Trichoderma asperellum* relieved plant oxidative damage while decreasing stress indicator enzyme levels such as malondialdehyde and peroxidase by 7-85% and 7-49%, respectively. Pb was primarily accumulated in plant roots and hardly translocated, according to the examination of phytoextraction characteristics and bioavailability. This suggests that phyto-stabilization acted as a critical healing mechanism. Conversely, the concentration of Na^+ and Ca^{2+} ions in plants was found in the following order: shoot > root. Furthermore, plant Pb concentrations and translocations decreased by 9-42%, 13-58%, and 19-30%, respectively, when *Trichoderma asperellum* was added to planted soil. Additionally, soil Pb bioavailability decreased by 6-21%. Thus, demonstrating phytostabilisation as a major remediation pathway.

Additional research by Li et al. (2018, 2020) revealed that *Enterobacter* sp. FM-1 may enhance Mn and Cd accumulation in plants by considerably raising their bioavailability by lowering soil pH. It can also stimulate the development of three different plants. Additionally, introducing *Serratia* PRE01 or *Arthrobacter* PRE05 into rhizosphere soil may improve the availability of organic matter for HMs, hence boosting phytoremediation of vanadium (V)-contaminated soil without causing harm to the environment (Wang et al., 2020). Thus, they can enhance phytoremediation efficiency, as given in Table 3.

Research articles on microbial-assisted phytoremediation provide evidence of the increase in bioaccumulation efficiency of heavy metals when different microbes are introduced into the soil rhizosphere of different plants, as given below.

Table 3: Impact of microbial assistance on phytoremediation efficiency

Plants	*Microorganisms*	*Metals*	*Effects*	*References*
Brassica juncea	*Serratia* PRE01, *Arthrobacter* PRE05	V	There was an increase in biomass (12.0-17.4%) and metal absorption (24.2-32.0%) in the inoculated treatments.	Wang et al. (2020)
Brassica juncea	*Sreptomyces pactum* Act12	Cd, Pb, Zn, and Cu	Zn (40, 14%) and Pb (82, 15%) were more readily absorbed by the shoot and root, respectively. There was a 17 and 33% increase in the root's absorption of Cd and Cu, respectively.	Ali et al. (2017)
Brassica oxyrrhina	*Pseudomonas libanensis* TR1, *Pseudomonas reactans* Ph3R3	Cu, Zn	In comparison to the control group, TRI enhanced the Cu concentration in *B. oxyrrhina* by 61%.	Ma et al. (2016)
Centena asiatica L.	*Enterobacter* sp. FM-I	Cd	Plants infected with *Enterobacter* sp. FM-I had Cd concentrations 71.5, 162.3 and 120.4% greater than the control's in their roots, stems and leaves, respectively.	Li et al. (2018)
Helianthus annuus	*Trichoderma* sp.	Pb, As	*Trichoderma* sp. MG was discovered to have branches with the highest As (67%) and Pb (59%) accumulations.	Govarthanan et al. (2018)
Helianthus annuus	*Pseudomonas libanensis* TR1, *Claroideoglom-us claroideum* BEG210	Ni	Introducing the strains into *Helianthus annuus* resulted in a 38-82% increase in Ni accumulation.	Ma et al. (2019)

(continued)

Table 3: (continued)

Plants	*Microorganisms*	*Metals*	*Effects*	*References*
Napier grass	*Micrococcus* sp.	Cd	Following *Micrococcus* sp. inoculation, Cd deposition in the roots and shoots rose 1.5 and 1.6 times, respectively. Introduction of *Arthrobacter* sp. resulted in a 1.4- and 1.5-fold increase in Cd accumulation in the roots and shoots.	Wiangkham and Prapagdee (2018)
Phragmites communis	*Simplicillium chinense* QD10	Cd, Pb	With the help of *S. chinense* QD10, the concentration of Cd in roots rose by 100.4%. In shoots, *S. chinense* QD10 helped raise the Cd content to 2 g kg^{-1}, a 243% rise.	Jin et al. (2019)
Sedum plumbizincicola	*Rhodococcus* sp. NSX2	Cd	After the NSX2 inoculation, Cd in the shoot rose by 21.7%.	Hou et al. (2019)
Sesbania rostrata	Rhizobia and arbuscular mycorrhiza fungi	U	Under plant-arbuscular mycorrhiza fungus rhizobia treatment, the U removal rates were 50.5-73.2%, but in plant-treated soil, the removal rates were only 7.2-23.3%	Ren et al. (2019)
Trifolium arvense	*Pseudomonas azotoformans* ASS1	Cu, Zn, Ni	ASSI raised the accumulations of Ni, Cd, and Zn by 26, 51 and 63%, respectively, in comparison to controls.	Ma et al. (2017)
Scirpus triqueter	Plant-growth-promoting rhizobacteria (PGPR	Ni	Ni removal rates were raised from 0.9 to 8.8% (almost nine times) by PGPR-inoculated S. triqueter.	Zhang et al. (2020a, 2020b)

4 Conclusion

Phytoremediation has been found to be a promising plant-mediated remediation technology. Although using traditional plants alone in bioremediation produces a less effective result, this problem can be addressed by using microbial-assisted phytoremediation and adding addenda like biochar and methods like genetic engineering to mixing. Even though much work has been done, the full potential of this technique is yet to be realised. Hence, further investment and research should be made in this regard. However, we need to work more on effective disposal techniques for contaminated biomass. This way, we can pave the road to a grand-scale utilisation of bioremediation technology from a laboratory.

References

Ali A, Guo D, Mahar A, Wang Z, Muhammad D, Li R, Wang P, Shen F, Xue Q, Zhang Z (2017) Role of *Streptomyces pactum* in phytoremediation of trace elements by *Brassica juncea* in mine-polluted soils. Ecotoxicol Environ Saf 144:387–395

Ali H, Khan E, Sajad MA (2013) Phytoremediation of heavy metals—concepts and applications. Chemosphere 91(7):869–881

Al-Baldawi IA., Abdullah SRS, Anuar N, Hasan HA (2018) Phytotransformation of methylene blue from water using aquatic plant (Azolla pinnata). Environ Technol & Innov 11:15–22

Bastos E, Schneider M, de Quadros DP, Welz B, Batista MB, Horta PA, Rörig LR, Barufi JB (2019) Phytoremediation potential of Ulva ohnoi (Chlorophyta): influence of temperature and salinity on the uptake efficiency and toxicity of cadmium. Ecotoxicol Environ Saf 174:334–343

Chen X, Wang J, Hayat K, Zhang D, Zhou P (2021) Small structures with big impact: multi-walled carbon nanotubes enhanced remediation efficiency in hyperaccumulator *Solanum nigrum* L. under cadmium and arsenic stress. Chemosphere 276:130130

Covre WP, da Silveira Pereira WV, Gonçalves DAM, Teixeira OMM, do Amarante CB, Fernandes AR (2020) Phytoremediation potential of Khaya ivorensis and Cedrela fissilis in copper contaminated soil. J Environ Manag 268:110733

de Souza TD, Borges AC, Braga AF, Veloso RW, de Matos AT (2019) Phytoremediation of arsenic-contaminated water by Lemna Valdiviana: an optimisation study. Chemosphere 234:402–408

Eze VC, Harvey AP (2018) Extractive recovery and valorization of arsenic from contaminated soil through phytoremediation using Pteris cretica. Chemosphere 208:484–492

Fayiga AO, Saha UK (2016) Soil pollution at outdoor shooting ranges: health effects, bioavailability, and best management practises. Environ Pollut 216:135–145

Ghazaryan KA, Movsesyan HS, Minkina TM, Sushkova SN, Rajput VD (2021) The identification of the phytoextraction potential of Melilotus officinalis and Amaranthus retroflexus growing on copper- and molybdenum-polluted soils. Environ Geochem Health 43(4):1327–1335

Ghosh M, Singh SP (2005) A review on phytoremediation of heavy metals and utilization of it's by products. Asian J Energy Environ 6(4):18

Govarthanan M, Mythili R, Selvankumar T, Kamala-Kannan S, Kim H (2018) Myco-phytoremediation of arsenic- and sp., lead-contaminated soils by Helianthus annuus and wood rot fungi, Trichoderma sp. isolated from decayed wood. Ecotoxicol Environ Saf 151:279–284

Guarino F, Ruiz KB, Castiglione S, Cicatelli A, Biondi S (2020) The combined effect of Cr (III) and NaCl determines changes in metal uptake, nutrient content, and gene expression in quinoa (Chenopodium quinoa Willd.). Ecotoxicol Environ Saf 193:110345

He L, Zhu Q, Wang Y, Chen C, He M, Tan F (2021) Irrigating digestate to improve cadmium phytoremediation potential of Pennisetum hybridum. Chemosphere 279:130592

Hou J, Liu W, Wu L, Ge Y, Hu P, Li Z, Christie P (2019) Rhodococcus sp. NSX2 modulates the phytoremediation efficiency of a trace metal-contaminated soil by reshaping the rhizosphere microbiome. Appl Soil Ecol 133:62–69

Javed MT, Tanwir K, Akram MS, Shahid M, Niazi NK, Lindberg S (2019) Phytoremediation of cadmium-polluted water/sediment by aquatic macrophytes: role of plant-induced pH changes. Cadmium Toxic Toler Plants 495–529

Jiang M, Liu S, Li Y, Li X, Luo Z, Song H, Chen Q (2019) EDTA-facilitated toxic tolerance, absorption, and translocation, and phytoremediation of lead by dwarf bamboos. Ecotoxicol Environ Saf 170:502–512

Jin Z, Deng S, Wen Y, Jin Y, Pan L, Zhang Y, Black T, Jones KC, Zhang H, Zhang D (2019) Application of Simplicillium chinense for Cd and Pb biosorption and enhancing heavy metal phytoremediation in soils. Sci Total Environ 697:134148

Li X, Zhang X, Wang X, Yang X, Cui Z (2019) Bioaugmentation-assisted phytoremediation of lead and salinity-contaminated soil by Suaeda salsa and Trichoderma asperellum. Chemosphere 224:716–725

Li Y, Liu K, Wang Y, Zhou Z, Chen C, Ye P, Yu F (2018) Improvement of cadmium phytoremediation by Centella asiatica L. after soil inoculation with cadmium-resistant Enterobacter sp. FM-1. Chemosphere 202:280–288

Li Y, Lin J, Huang Y, Yao Y, Wang X, Liu C, Liang Y, Liu K, Yu F (2020) Bioaugmentation-assisted phytoremediation of manganese and cadmium-contaminated soil by Polygonaceae plants (Polygonum hydropiper L. and Polygonum lapathifolium L.) and Enterobacter sp. FM-1. Plant Soil 448(1):439–453

Li Y, Xie T, Zha Y, Du W, Yin Y, Guo H (2021) Urea-enhanced phytoremediation of cadmium with willow in pyrene and cadmium-contaminated soil. J Hazard Mater 405:124257

Lim KT, Shukor MY, Wasoh H (2014) Physical, chemical, and biological methods for the removal of arsenic compounds. BioMed Res Int

Lin H, Zhang XH, Chen J, Liang L, Liu LH (2019) Phytoremediation potential of Leersia hexandra Swartz in copper-contaminated soil and its enhancement by using agronomic management practises. Ecol Eng 127:561–561

Liu X, Mao Y, Zhang X, Gu P, Niu Y, Chen X (2019) Effects of PASP/NTA and TS on the phytoremediation of pyrene-nickel-contaminated soil by Bidens pilosa L. Chemosphere 237:124502

Liu ZQ, Li HL, Zeng XJ, Lu C, Fu JY, Guo LJ, Kimani WM, Yan HL, He ZY, Hao HQ, Jing HC (2020) Coupling phytoremediation of cadmium-contaminated soil with safe crop production based on a sorghum farming system. J Clean Prod 275:123002

Liu Z, Chen B, Wang LA, Urbanovich O, Nagorskaya L, Li X, Tang L (2020) A review on phytoremediation of mercury-contaminated soils. J Hazard Mater 400:123138

Ma Y, Rajkumar M, Moreno A, Zhang C, Freitas H (2017) Serpentine endophytic bacterium Pseudomonas azotoformans ASS1 accelerates phytoremediation of soil metals under drought stress. Chemosphere 185:75–85

Ma Y, Rajkumar M, Oliveira RS, Zhang C, Freitas H (2019) Potential of plant-beneficial bacteria and arbuscular mycorrhizal fungi in phytoremediation of metal-contaminated saline soils. J Hazard Mater 379:120813

Ma Y, Rajkumar M, Zhang C, Freitas H (2016) Inoculation of Brassica oxyrrhina with plant growth-promoting bacteria for the improvement of heavy metal phytoremediation under drought conditions. J Hazard Mater 320:36–44

Manori S, Shah V, Soni V, Dutta K, Daverey A (2021) Phytoremediation of cadmium-contaminated soil by Bidens pilosa L.: impact of pine needle biochar amendment. Environ Sci Pollut Res 28(42):58872–58884

Meharg AA, Cairney JW (2000) Ectomycorrhizas—extending the capabilities of rhizosphere remediation? Soil Biol Biochem 32(11–12):1475–1484

Mendoza-Hernández JC, Vázquez-Delgado OR, Castillo-Morales M, Varela-Caselis JL, Santamaría-Juárez JD, Olivares-Xometl O, ... Pérez-Osorio G (2019) Phytoremediation of mine tailings by Brassica juncea inoculated with plant growth-promoting bacteria. Microbiol Res 228:126308

Minisha MT, Shah IK, Varghese GK, Kaushal RK (2021) Application of Aztec Marigold (Tagetes erecta L.) for phytoremediation of heavy metal polluted lateritic soil. Environ Chem Ecotoxicol 3:17–22

Mirzaei M, Verrelst J, Bakhtiari AR, Marofi S (2021) Potential use of grapevine cv. Askari for heavy metal phytoremediation purposes at greenhouse scale. Environ Sci Pollut Res 28(10):12447–12458

Nwoko CO (2010) Trends in phytoremediation of toxic elemental and organic pollutants. Afr J Biotechnol 9(37):6010–6016

Odokuma LO, Akponah E (2010) Effect of concentration and contact time on heavy metal uptake by three bacterial isolates. J Environ Chem Ecotoxicol 2(6):84–97

Padmavathiamma PK, Li LY (2007) Phytoremediation technology: hyper-accumulation metals in plants. Water, Air, Soil Pollut 184:105–126

Qin Y, Shi X, Wang Z, Pei C, Cao M, Luo J (2021) Influence of planting density on the phytoremediation efficiency of Festuca arundinacea in Cd-polluted soil. Bull Environ Contam Toxicol 107(1):154–159

Ren CG, Kong CC, Wang SX, Xie ZH (2019) Enhanced phytoremediation of uranium-contaminated soils by arbuscular mycorrhiza and rhizobium. Chemosphere 217:773–779

Sarma H (2011) Metal hyperaccumulation in plants: a review focusing on phytoremediation technology. J Environ Sci Technol 4(2):118–138

Shah V, Daverey A (2020) Phytoremediation: A multidisciplinary approach to cleaning up heavy metal-contaminated soil. Environ Technol Innov 18:100774

Sharma S, Singh B, Manchanda VK (2015) Phytoremediation: role of terrestrial plants and aquatic macrophytes in the remediation of radionuclides and heavy metal contaminated soil and water. Environ Sci Pollut Res 22:946–962

Shi L, Deng X, Yang Y, Jia Q, Wang C, Shen Z, Chen Y (2019) A Cr (VI)-tolerant strain, Pisolithus sp1, with a high accumulation capacity of Cr in mycelium and highly efficient assisting Pinus thunbergii for phytoremediation. Chemosphere 224:862–872

Sidhu, G.P.S., Bali, A.S., Singh, H.P., Batish, D.R. and Kohli, R.K. (2020). Insights into the tolerance and phytoremediation potential of Coronopus didymus L. (Sm) grown under zinc stress. *Chemosphere*, **244:** 125350.

Tang L, Hamid Y, Zehra A, Sahito ZA, He Z, Hussain B, Gurajala HK, Yang X (2019) Characterisation of fava bean (Vicia faba L.) genotypes for phytoremediation of cadmium and lead co-contaminated soils coupled with agro-production. Ecotoxicol Environ Saf 171:190–198

Wang J, Feng X, Anderson CW, Xing Y, Shang L (2012) Remediation of mercury contaminated sites – a review. J Hazard Mater 221:1 18

Wang L, Lin H, Dong Y, Li B, He Y (2020) Effects of endophytes inoculation on rhizosphere and endosphere microecology of Indian mustard (Brassica juncea) grown in vanadium-contaminated soil and its enhancement on phytoremediation. Chemosphere 240:124891

Wei S, da Silva JAT, Zhou Q (2008) Agro-improving method of phytoextracting heavy metal contaminated soil. J Hazard Mater 150(3):662–668

Wetle R, Bensko-Tarsitano B, Johnson K, Sweat KG, Cahill T (2020) Uptake of uranium into desert plants in an abandoned uranium mine and its implications for phytostabilization strategies. J Environ Radioact 220:106293

Wiangkham N, Prapagdee B (2018) Potential of Napier grass with cadmium-resistant bacterial inoculation on cadmium phytoremediation and its possibility to use as biomass fuel. Chemosphere 201:511–518

Willscher S, Jablonski L, Fona Z, Rahmi R, Wittig J (2017) Phytoremediation experiments with Helianthus tuberosus under different pH and heavy metal soil concentrations. Hydrometallurgy 168:153–158

Wu G, Kang H, Zhang X, Shao H, Chu L, Ruan C (2010) A critical review on the bio-removal of hazardous heavy metals from contaminated soils: Issues, progress, eco-environmental concerns and opportunities. J Hazard Mater 174(1–3):1–8

Yang C, Ho YN, Inoue C, Chien MF (2020) Long-term effectiveness of microbe-assisted arsenic phytoremediation by Pteris vittata in field trials. Sci Total Environ 740:140137

Zeng G, Wan J, Huang D, Hu L, Huang C, Cheng M, ... Jiang D (2017) Precipitation, adsorption and rhizosphere effect: the mechanisms for phosphate-induced Pb immobilization in soils—a review. J Hazard Mater 339:354–367

Zhang X, Su C, Liu X, Liu Z, Liang X, Zhang Y, Feng Y (2020) Effect of plant-growth-promoting rhizobacteria on phytoremediation efficiency of Scirpus triqueter in pyrene-Ni co-contaminated soils. Chemosphere 241:125027

Zhang Y, Chen Z, Xu W, Liao Q, Zhang H, Hao S, Chen S (2020) Pyrolysis of various phytoremediation residues for biochars: chemical forms and environmental risk of Cd in biochar. Biores Technol 299:122581

Zhao X, Liu W, Cai Z, Han B, Qian T, Zhao D (2016) An overview of preparation and applications of stabilised zero-valent iron nanoparticles for soil and groundwater remediation. Water Res 100:245–266

Index

P. K. Padhy et al. (eds.), *Airborne Particulate Matter*,
https://doi.org/10.1007/978-3-031-84408-9

GPSR Compliance

The European Union's (EU) General Product Safety Regulation (GPSR) is a set of rules that requires consumer products to be safe and our obligations to ensure this.

If you have any concerns about our products, you can contact us on

ProductSafety@springernature.com

In case Publisher is established outside the EU, the EU authorized representative is:

Springer Nature Customer Service Center GmbH
Europaplatz 3
69115 Heidelberg, Germany

www.ingramcontent.com/pod-product-compliance
Lightning Source LLC
Chambersburg PA
CBHW051632080625
27894CB00001B/32
9783031844072